ACCLAIM FOR

TO TOUCH A
WILD DOLPHIN

"*To Touch a Wild Dolphin* is comparable to Jane Goodall's great classic *In the Shadow of Man*. . . . This book is undoubtedly important."　　　　　　　　**—*Daily News***

"Evocative . . . and thought-inspiring. . . . Fascinating work, part of the animal research that has led us to redefine intelligence and reconfigure our notion of kinship with other species."　　　　　　　　**—*Kirkus Reviews***

"An intimate, engaging glimpse into the world of wild dolphins. [Smolker's] worry for the safety of her cetacean acquaintances lends gravity to this animated, empathetic account of life among Flipper's wild kin."　　　　　　　　**—*Publishers Weekly***

"Smolker has the ability to take the reader with her as she wades into the water to visit the dolphins. . . . Entrancing reading."　　　　　　　　**—*Booklist***

"Lyrical prose . . . offering captivating insight into research in the field."　　　　　　　　**—*Science News***

Rachel Smolker

TO TOUCH A
WILD DOLPHIN

Rachel Smolker co-founded the Monkey Mia Dolphin
Research Project in 1982, which continues to produce
groundbreaking insights into virtually every aspect of
dolphin life. She has participated in other studies of
dolphins and whales all over the world, including
British Columbia, the Bahamas, and New Zealand.
She has also observed various species of primates in
Southeast Asia, Central America, and Madagascar. She
is currently a research associate at the University of
Vermont and maintains an affiliation with the Museum
of Zoology at the University of Michigan, where she
completed her doctorate. She lives in Vermont.

TO TOUCH A
WILD DOLPHIN

TO TOUCH A WILD DOLPHIN

A Journey of Discovery with the Sea's
Most Intelligent Creatures

RACHEL SMOLKER

ANCHOR BOOKS

A DIVISION OF RANDOM HOUSE, INC. • NEW YORK

FIRST ANCHOR BOOKS EDITION, JULY 2002

Copyright © 2001 by Rachel Smolker

All rights reserved under International and Pan-American Copyright
Conventions. Published in the United States by Anchor Books, a division
of Random House, Inc., New York, and simultaneously in Canada by
Random House of Canada Limited, Toronto. Originally published in hardcover
in the United States by Nan A. Talese, an imprint of Doubleday,
a division of Random House, Inc., New York, in 2001.

Anchor books and colophon are registered trademarks of Random House, Inc.

The Library of Congress has cataloged the Nan A. Talese/Doubleday edition as follows:
Smolker, Rachel.
To touch a wild dolphin / Rachel Smolker.—1st ed.
p. cm.
ISBN 0-385-49176-X
1. Smolker, Rachel. 2. Biologists—United States—Biography.
3. Dolphins—Behavior—Australia—Shark Bay (W.A.) I. Title.
QH31.S6124 A3 2001
599.53'15—dc21
[B]
00-053649

Anchor ISBN: 0-385-49177-8

Book design by Deborah Kerner/Dancing Bears Design
Illustrations of dolphins by Richard Waxberg/Dancing Bears Design

www.anchorbooks.com

Printed in the United States of America
10 9 8 7 6 5 4 3 2 1

TO THEM AND US

Contents

TO TOUCH A
WILD DOLPHIN

PROLOGUE

A loud, percussive *pfhooo* awakens me from the shallow half-sleep one has on board a boat. I lay still for a moment, wide-eyed, listening. No doubt a dolphin breathing and, again, surprisingly close. After pulling back the salt-damp covers, I clamber out onto the deck of *Nortrek*, our forty-foot catamaran. A persistent cool breeze blows out of the southeast, and the stars are blinking fiercely, a broad and brilliant arch overhead. The moon is trailed by a strip of shimmering water, but otherwise the water is dark and calm. The flow of an incoming tide tugs gently on *Nortrek*'s moorings and slips back along her twin hulls. There in the moonlight I see the silvery shape of a dolphin's back rolling at the surface as it breathes and submerges in one fluid motion. Then a burst of glittering phosphorescence shoots forward like a comet and dissipates in a sparkling splash as the dolphin lunges after a fish, then breaks through the surface to breathe again.

I can just barely make out the dorsal fin, squat with nicks along the top edge; it is Nicky. She moves past the line of moonshine, and her silver-smooth skin glows as she rolls back under the surface and is transformed into another comet of phosphorescence. I know by her breathing and the way she is moving that she is hunting.

My mind still in that floating, receptive state of the recently asleep, I settle down on the deck to admire the spectacle: the phosphorescent comets below and the Milky Way above. The magnificence of the scenery pulls me far above and beyond myself. Shark Bay is a tremendous, wide-open expanse, jutting out into the Indian Ocean. Distant from any city lights, it is a place where the night skies offer up a slowly rotating banquet of constellations, pulsating multicolor planets, bright clouds of star clusters, and dark, eerie nebulae. The occasional passing satellite and shooting star are the only objects that disrupt an otherwise constant and by now familiar geometry. Right now Orion is low on the horizon, so it must be about three A.M.

On a night like this, the water seems literally to seethe with creatures that flip, splash, jump, skitter, snap, and gulp at the surface. Just a small indication of the world that lurks below. I am glad to be sitting up above it on *Nortrek*'s deck, high and dry, and can't help but wonder what it is like for Nicky, moving through this underwater world in the dark. Her echolocation permits her to "see" things with sound, but only in a narrow flashlight beam in front of her. So many of the creatures that are swimming around with her in this dark liquid soup are dangerous, even deadly. This is Shark Bay, after all. Some of the sharks are harmless, but

To Touch a Wild Dolphin

others, like the tiger shark, prey on dolphins. There are also poisonous scorpionfish with their elaborately decorated and deadly spines, the hideous cryptic stonefish, whip-fast stingrays, and sea snakes. All are capable of making life miserable for a dolphin who might blunder into them on a dark night.

I hear at least one other dolphin breathing in the distance—probably Nicky's mother, Holeyfin, and perhaps also her age-mate and friend Puck. After listening to the rhythms of their breathing for a while longer, I can tell that Nicky is progressing out in the direction of the other dolphins. I can envision the others pausing to wait for her as she catches up: gliding alongside Holeyfin, Nicky subtly tilts her belly toward her mother in greeting as they both dive down to examine a sea-grass patch on the bottom where a fish has taken refuge. I know Nicky well, and she is not prone to excessive displays of affection. She is smart, usually quite serious, hard to please, and introspective, it appears, at times. In some ways she reminds me of myself, and I have always felt a particularly strong bond with her.

Nicky, her family, and the many other dolphins in Red Cliff Bay (a small embayment within the much larger Shark Bay) have been the focus of much of my life for over fifteen years now. I have been granted the privilege of sharing their world and in the process have cultivated a deep affection for them. It is the sort of eager affection I have felt at times when encountering some unusually interesting and exotic foreigners while traveling in their country. Though we can barely communicate, and I know little about their world, somehow the juxtaposition of what is common and what differs between us inspires in me both a deep sense of kinship and a keen and inescapable awareness of being merely an observer. With dolphins, the kinship is further removed, and I am even more an observer from outside.

Perched aboard this fragile shell of a boat, as she clings to her anchor against the shoreline of this remote, windswept outpost, I feel the vastness that surrounds me on all sides—it is a tangible sensation. To the west, across the entire width of the Indian Ocean, lies Madagascar and the east coast of Africa. Far to the north lies Indonesia, and to the south, just wind, waves, and ocean reach all the way to Antarctica. To the east lies the tiny fishing camp of Monkey Mia and, beyond that, hundreds of miles of desert, the outback of Australia. Above, the unfathomable expanse of space.

This perspective enhances my sense of kinship with Nicky and the other dolphins. After more than fifteen years spent observing them, I, along with several other members of our research team, have made many discoveries about the lives of the Shark Bay dolphins. More than anything, what we have learned has defined the contours of what we do not know about them. Yet in spite of all the senses, experiences, and capabilities that we cannot share with these dolphins and do not understand, there is a simple, vital, powerful, warm-blooded commonality that binds us.

My journey into the world of the Shark Bay dolphins really began when I was a child, although I didn't realize it at the time. Most dolphin enthusiasts I've talked with claim their interest in dolphins was sparked during childhood, usually from watching *Flipper* reruns on television. I never watched *Flipper* and really never gave dolphins much thought. But I was the daughter of two biologists and was obsessed with animals from an early age. My father, an ornithologist, was the local repository for baby birds fallen from nests or injured birds discovered by our neighbors on rural Long Island. These he handed on to me to take care of. I spent all my free time searching for inchworms and grasshoppers, coaxing the babies to eat, keeping them warm and clean.

Sorting through a box of my mother's old photographs recently, I found a collection of shots labeled "Rachel and Her Pets." In one photograph, a beautiful yellow-shafted flicker is perched on my hand. I look both thrilled and a bit nervous, as though the responsibility of posing with the bird for my father's camera were an onerous one. This flicker had come to us as a baby, and I spent most of a summer diligently carrying it through the woods, seeking out termite nests and teaching it to forage. Later he (or she?) would follow me through the woods as I rode on horseback, returning occasionally to perch on my shoulder and probing my ears and nostrils and eyes with that phenomenally long woodpecker's tongue. There is also a series of pictures of me with various baby ducks. Imprinted from hatching, they followed me everywhere, and I adored them.

The birds were part of our family's larger menagerie: dogs, cats, a horse, guinea pigs, rabbits, raccoons, gerbils, hamsters, turtles, chameleons, snakes, a toadfish. In my child's mind, the greatest pleasure of all was to watch and care for animals. I loved them with an intensity that I later recognized only with the birth of my own children. Their well-being, safety, and comfort were everything to me.

I believe now that those experiences with animals were not by any means trivial to my development as a human being. They taught me about the responsibility involved in taking care of another living creature: I remember sleeping night after night with a baby bluejay carefully cupped in my hands for warmth. They taught me about how life unfolds: I remember being completely awestruck when my mother and I made tiny windows into the shells of incubator-raised chicken eggs using wax and microscope slides so we could watch the chicks develop within. And they taught me about death: when my pets died or disappeared, as they all did sooner or later, I was distraught but learned to cope.

As an adolescent, my interests turned from animals to boys (perhaps not so different). I followed one of those boys, Peter Birns, to California after dropping out of high school. Peter got me interested in dolphins; together we read John Lilly's books about dolphins: *Man and*

Dolphin; The Mind of the Dolphin; and *Communication Between Man and Dolphin.* Lilly, a neurobiologist by training, had brought to public attention the fact that dolphins have extraordinarily large brains. He claimed that dolphins might be even more intelligent than humans and that they might have a language on par with our own.

Unfortunately, as Lilly became more and more enamored of experimentation with LSD, his arguments on behalf of dolphin brainpower became increasingly incoherent. Nonetheless, his books planted the notion in many minds, including my own, that there might just be another species on the planet—in the oceans—possessed of a sentient intelligence and capable of tremendous empathy and kindness. Fellow earth travelers with whom we might commune intelligently and who might be able to relate to our deepest thoughts, ideas, and yearnings. These ideas were fantastic, inspiring, compelling.

Neither Peter nor I had ever seen a real live dolphin, but we were convinced that dolphins are highly intelligent. Magically so. Not only that, but they are also gentle, perpetually smiling, and kindly disposed toward people.

After all, like most people, we had heard stories about dolphins rescuing people in trouble at sea, befriending children, and helping fishermen. We had seen the old movie classic *Day of the Dolphin,* where the dolphin star engages in heroic antics to rescue George C. Scott from various dangerous situations, all the while reciting "Fa loves Pa" in a dolphin imitation of human speech.

Some of us have had the good fortune to see dolphins in the wild, but mostly we see them on television (*Flipper*) or in the movies (*Free Willy*). At best we might have the opportunity to watch dolphins perform in oceanaria like Sea World, where they seem to enjoy entertaining us with breathtaking leaps, friendly antics, and perpetual smiles. From these sources, part fact and part fantasy, most people derive a sense that dolphins are sweet, lovable, friendly, and very smart. Peter and I were taken in. What could be more appealing than an animal both highly intelligent and also good-natured? My interest in dolphins was sparked before I had ever even met one face-to-face.

My first opportunity to do so came about in 1980. Following Peter around California from college to college as he pursued a degree in psychology, I had kept myself busy with odd jobs, making music and pottery. But when Peter transferred to UC Santa Cruz, I finally came full circle and returned to my roots as a biologist. UCSC, fondly known as "Uncle Charlie's Summer Camp," was also home of the Long Marine Laboratory, where Ken Norris, a pioneer in the study of dolphin biology, had an office. I simply had to be his student, so I crammed as many credits as was humanly possible into one semester at the local community college in order to meet the requirements to transfer to UCSC.

Ken invited me to participate in his research project, studying the Hawaiian spinner dolphins. These dolphins, named for their spectacular spinning leaps, came every morning into Kealakekua Bay, on the Kona coast of the big island of Hawaii, where they would rest and play during the day. Late in the afternoon they would begin to rouse, their aerial acrobatics becoming more and more athletic as they gained momentum for a grand exodus out to sea, where they spent their nights foraging in deep, dark waters.

My year with the spinner dolphins was a fantasy come true, but it was also frustrating. Spending my days watching dolphins, living in a tropical paradise, eating mangos and papayas and fresh fish, swimming in clear, coral-filled waters, was idyllic. We watched the dolphins from shore, from boats, from a clifftop surveyors theodolite station, from a small airplane, and in an aquarium, the Sea Life Park on Oahu. But rarely were we able to identify the dolphins as individuals even though we took hundreds of photographs of the natural markings on their dorsal fins in an attempt to learn to do so. These fast, sleek, oceanic dolphins seemed to operate on a time scale much faster than ours. We caught fleeting glimpses of their lives, and more often than not, things happened so fast as to be incomprehensible to us. At the end of a year I had learned a lot about how to go about watching dolphins, but I didn't get to know them in the ways that I craved. It was as though I had tried to satisfy an interest in people by making a study of subway riders in New York City rather than making friends with a few individuals. I wanted to know a dolphin personally. I wanted to put my hand against its skin, look it in the eye, and develop and explore some sort of more personal rapport.

Back at UCSC, I found myself sitting in on graduate seminars on the evolution of social behavior with Bob Trivers, who is responsible for several of the most important theories about why animals (including humans) behave as they do. Drawing upon an awesomely broad knowledge of biology, Bob gave lectures that ranged from the ridiculous to the sublime. When he was on, there was nothing to do but sit back and concentrate hard in order to keep up as he led you through his reasoning to new and mind-expanding insights. It was while reading one of his papers, "The Evolution of Reciprocal Altruism," that I experienced a life-changing epiphany. My grasp of evolutionary theory had reached a gel point, and suddenly everything, from the behavior of ants, bees, and wasps to the way I felt about certain people in my own personal life, made sense to me.

Anyone who finds it impossible to ignore the bones and stones of the fossil record believes in evolution. For a few creatures, the fossil record is so well documented that we can easily trace how each successive species differed ever so slightly, transforming from an ancient form to another, perhaps quite different modern one.

Roughly sixty million years ago, a creature that was also the ancestor to the ungulates (cows, hippopotami, and their two-toed relatives) began a gradual return to the sea. Initially foraging in the shallows, a bit like modern raccoons and otters, these newly aquatic mammals were probably not too adept as swimmers. But there were advantages to venturing into deeper and deeper waters, and over time they became more and more aquatic. Their nostrils moved toward the top of their heads, where the modern dolphin's blowhole now lies, making it easier to surface and breathe. Their ears became modified to take advantage of underwater sounds. Their eyes adapted to seeing underwater. They gradually lost their hind limbs and evolved a torpedolike shape powered along by strong tail flukes. Their forelimbs became fused into flippers, and they lost their fur. Even their skin changed, to withstand the constant soaking. Virtually every aspect of their physical form and physiology had changed dramatically, adapting to the specific challenges of a marine existence.

This was a gradual process, and one that proceeded along without any conscious plan or goal in mind. Instead, over countless generations, the few rare mutations in those ancestral cetaceans that helped the animals better to cope with the marine environment also enabled them to leave more offspring. In the parlance of evolutionary biology, those individuals better adapted to a marine lifestyle experienced greater reproductive success. In the next generation, more individuals carried those genes and those characteristics. Biologists, beginning with Darwin, refer to this process of gradual modification as "natural selection."

Though selection is seemingly a simple concept, the ramifications are tremendous. Perhaps the most significant is that the "currency" of evolution is reproductive success. Reproduce well (meaning more or better than others of your species), and you shall be represented in future generations. Fail, and you contribute nothing toward shaping the future of your kind.

It was, for me, a revelation to finally understand that the most important and defining feature of the behavior and biology of all animals, including humans, was reproduction. I hadn't thought much about reproducing myself at that time, and few people around me at college were talking about it. Certainly none were scheming about ways to ensure their genetic representation in future generations.

But there is no need for any organism to be aware of the mechanisms of evolution. A spider does not know why she spins a web, but she does so anyway, and does a good job of it, because her ancestors made better webs and thereby outreproduced their spiderly neighbors. The importance of reproduction and the drive to do so may be completely hidden from our conscious consideration. Instead, a million other seemingly unrelated little matters may occupy our thoughts and direct our actions. Yet on closer analysis, indirect and circuitous though the path may be, it is the impact of those little matters on reproduction that shapes the future course of individuals and species.

The more I read and thought about evolution, the more I found myself thinking in terms of an economy of behavior, with the currency being costs and benefits to the goal of reproduction. A simple implication of the mechanism of evolution is that any individual should, for the most part, only do things that will help him or her to reproduce and avoid incurring costs.

An animal who generously hands over all his food to another will likely starve to death and fail to leave any offspring. Any genes that played a role in this generous behavior will go extinct with him, while an animal who hordes as much food as possible and is able to fatten himself up, feed his offspring well, and leave some to spare for an emergency will leave more surviving offspring than his generous counterpart, and the genes involved in hoarding behavior will be passed on. Thus, behaviors like generosity are somewhat surprising and quite rare.

The logic of evolution is not always "nice" according to our moral standards. But all behaviors involve a complex interplay among many genes and the environmental context in which they are expressed, so the process is not as neat and tidy as it sounds. One of the fantastic characteristics of human beings is that our behaviors are not bound by our history or our genetic makeup. Both the most hideous, selfish behaviors—warfare, thievery, infanticide—and the most glorious, selfless behaviors, like friendship, generosity, love, and intelligence, spring from the same basic process: adaptation through differential reproductive success. Base and heartless though the process of natural selection may seem, it is possessed of a distinct elegance.

With my newfound comprehension of how evolution works came some exciting new questions about dolphins, particularly dolphin intelligence. The dolphin brain is the largest (relative to body size) of any animal other than humans. Having a large brain is quite a remarkable characteristic. Brains are extremely expensive organs, requiring a huge amount of caloric energy to operate. Animals with large, expensive brains must gain some major advantage or they simply would not maintain such a "gas guzzler." The question is, what are the advantages, and in what circumstances do the benefits of such an organ outweigh the costs?

Physiologically, dolphin brains differ from human brains in some significant ways. For example, the cortex is the part of the brain involved in such "higher-order processes" as abstract thought and reasoning. It is considered to be the seat of consciousness. It is also the most recent innovation in human brain evolution. Around one million years ago the ancestral human skull began an exponential expansion in size, apparently to accommodate an expansion of the cortex, roughly coincident with the origins of design and construction of fine tools, art, complex civilizations, and distinctive cultures—the aspects of our own behavior that we consider to be hallmarks of human intelligence.

Dolphins also have a very large cortex, but it is spread more thinly over the rest of the brain, and the neurons that make up the cortex differ in form and organization from those in the human cortex. What this means about how dolphins think and feel is hard to say. Do they have the same powers of reasoning that we have? Are they self-aware? Do they feel love and hate, compassion, trust, distrust? Do they wonder about death? Do they have ideas about right and wrong and accompanying feelings of guilt and righteousness? What could they teach us about the oceans? How do they feel about one another? What do they think of us?

Finding out what dolphins do with their big and very different brains is no simple task. It is hard enough to figure out what might go on in the brains of our fellow humans. But certainly the most frequently asked question about dolphins is "How smart are they?" We think of ourselves as being smart. Does that mean that dolphins are smart only insofar as they are like us? Why not reverse the argument and compare us to them? We would probably come out looking pretty foolish, but who's to say what the standard should be?

Many a biologist and psychologist has struggled with the difficult question of defining intelligence. The more we have looked and conducted experiments and compared different species, the more we have been forced to rephrase the question "How smart are they?" to ask "How are they smart?"

Scientists studying animal intelligence have come to think in terms of the learning specializations that different types of animals exhibit. Each creature seems to be able to learn certain things and not others, and what it can learn may depend partly on the nature of the problem, but also on the particular stimuli involved. Some animals have shown us that they are highly capable of learning complex skills, but only under very precise conditions.

John Garcia's classic experiments with rats are a fine example of learning specializations. He was interested in how animals learn taste aversions. He easily trained rats to associate certain tastes with feelings of nausea. The rats quickly learned to avoid flavors that had previously led to nausea, and they made these associations even when there was a long time delay between the taste and the nausea. He also trained them to associate certain sounds with unpleasant skin sensations (electric shocks). But he found it extremely difficult to train the rats to associate taste with shock. The implication is that animals like rats are predisposed to associate tastes with nausea, but ill disposed toward developing associations between taste and shock. This makes sense given that in their natural environment rats must learn to avoid dangerous foods. Learning to associate one stimulus with another depends on the relationship between the two stimuli and on the biology of the animal. Developing good tests of an animal's intelligence therefore requires a good solid understanding of the animals' natural history. And since different species have very different natural histories, it is hard to come up with one universally applicable test that we can use to compare many different creatures.

Certainly our best measure of dolphin intelligence comes from watching what they do rather than looking at their brain anatomy and guessing how they might use such an organ. We can study dolphins in laboratories, devising tests that we think might reveal "how they are smart," but ultimately we must also go out on the oceans and see what they do, how they live. It is out there, after all, that they evolved those large brains over millions of years, and it is there that they now make use of them.

Questions about how dolphins use their brains in the wild are compelling because of what we might learn about dolphins, but also because, in the process, we might learn something about our own intelligence. Dolphins avail us of an invaluable opportunity: because they are so different from humans and yet, like us, they have evolved very large brains, they provide an especially fascinating point of comparison. We can ask, "What is a big brain good for? Under what circumstances do big brains evolve? What is the relationship between brain size and intelligence?" We live in such incomprehensibly different worlds, what could possibly have led, in two such unrelated creatures, to the evolution of big brains and intelligence? Could both dolphins and humans have been faced with challenges in adapting to their very different ecological circumstances that are somehow parallel? Perhaps there is something about the way we search for food, avoid predators, navigate through our surroundings, or deal with social relationships that is similar.

Questions about dolphin intelligence, and the evolution of intelligence in general, became the focus of my intellectual motivation for studying dolphins, but what really endeared them to me were a few stories I came across from people who had the opportunity to interact with dolphins face-to-face.

One such story that made a particularly strong impression involved a baby dolphin, housed in a large tank at the Port Elizabeth aquarium. She and her mother and the other dolphins in their tank were the subject of some of the first detailed observations of dolphin behavior. One day, a researcher was sitting by the glass observation window, smoking a cigarette. The baby dolphin came to the window, taking an interest in the smoke rising from the cigarette. As the smoke trailed up along the glass, the baby followed it intently. Then she swam over to her mother, suckled for a moment, came back to the window, and released a trickle of milk from the corner of her mouth. The milk rose against the glass, just as the cigarette smoke had done.

This story and others like it offered rare glimpses into the creativity and intelligence of dolphins and inspired me to embark upon a journey into the dolphin's world. Over the course of more than fifteen years spent watching wild dolphins, my colleagues and I would make discoveries that profoundly changed the way people think of dolphins, the way we define intelligence, and the sense of kinship we feel with another species.

More personally, this journey into the dolphin's world taught me not only about them, but about myself and my fellow humans. It is as if, having for so long strained to see dolphins moving below the reflective surface of the water, my eyes suddenly shifted their depth of focus and I realized that I was all along staring through my own reflection.

GETTING TO

MONKEY

MIA

I n 1981 Elizabeth Gawain was passing through Santa Cruz. A retired city planner who now spent much of her time traveling the world, teaching the occasional yoga class, she was a woman wise from her sixty-plus years, but she still retained the energy and curiosity of a child. Her unadorned eyes sparkled from beneath a stray hank of graying hair. Tiny laugh lines spread from the corners of her eyes and the edges of her mouth. She carried herself with the graceful ease of someone who had always been limber and strong and was not at all self-conscious about being unusual.

Elizabeth had just returned from a visit to a place called Monkey Mia in western Australia, where, quite remarkably, wild dolphins came directly into a shallow, sandy beach and accepted fish offered to them right out of people's hands. Though not a trained scientist, Elizabeth was a keen observer, and she recognized instantly the potential for learning about the lives of these friendly dolphins. She came to Long Marine Laboratory to tell Ken Norris about the Monkey Mia dolphins and presented a small talk and slide show to Ken and his band of students. She told us what she knew about the lives of each of the dolphins she had learned to identify: Holeyfin, Joy, Nicky, Crookedfin, Puck, Snubnose, and Bibi.

The dolphins were a well-kept secret, familiar to only a handful of people. Sometime during her presentation, Elizabeth mentioned that Monkey Mia would be an ideal place for a scientific study of some sort. The idea of traveling to a remote and exotic place in Australia and making friends with wild dolphins seemed too good to be possible.

After her talk, I joined a small crowd of eager dolphinologists as we walked Elizabeth out to the parking lot, trying to get in the last of our questions. Just before getting into her car, she paused. Focusing her attention squarely on me, she looked me in the eye with a penetrating gaze and said, "You'll go there." Then she got into her car and drove away. I was dumbfounded. It was not just the fact of her saying this, but the way in which she did so. As if she knew.

Afterward, as I rode my bicycle home, I tried to imagine what it would be like in fact to go to Australia. I wanted to learn about dolphins at least in part from interacting with them directly. I wanted to know something about their minds, their emotions, their relationships with each other, their personalities. I had begun to imagine that this would be possible only with captive dolphins. Wild dolphins were simply too inaccessible. They were difficult to find and too busy with their own lives to bother with me.

I had spent a little time watching captive dolphins at Steinhart Aquarium in San Francisco. The Steinhart dolphins were certainly interested in me. They would look me in the eye, and I did develop some sense of dolphin presence from my time with them. But they lived a bizarre and unnatural existence: two strangers thrown together in a tiny, virtually empty tank with one wall made of glass through which an endless stream of human visitors peered. I wanted to watch dolphins in the wild, but I also wanted to see more than the occasional fleeting glimpses we had experienced with the Hawaiian spinner dolphins.

At the time, there was one long-term study of wild dolphins under way, in Florida, under the direction of Randy Wells. The emphasis of Randy's studies, which had been going on for a number of years already, was population biology. He and his co-workers kept track of the movements of dolphins around Sarasota and regularly captured the dolphins for measurements and samples of tissues. Interesting, but not what I was looking for.

I wanted to be a part of the dolphin world, not force them to become part of mine. Monkey Mia seemed like the perfect solution: wild dolphins accustomed to interacting with people.

Even the most basic questions about dolphin biology, upon which any deeper understanding of their lives would necessarily be based, remained unanswered. How long do they live? How often do they have babies, and how are dolphin youngsters raised? Where and how far do they travel? What sorts of groups do they form? What do they eat, and how do they catch their prey? How much time do they spend at different activities? What are their social relationships like?

Considering the possibilities for discovery that the Monkey Mia dolphins might provide, my mind reeled with questions. That evening the phone rang. It was Richard Connor, a fellow student, who had also attended Elizabeth's talk that afternoon. I didn't know Richard very well at that time. We had been in a few classes together, and I knew he was intensely bright. He impressed me as something of a big, sloppy puppy dog, exuding a boyish enthusiasm for all things dolphin. "Why don't we just go there and check it out," he suggested. That was all the encouragement I needed. Richard and I talked to our various mentors: Ken Norris, Bernd Wursig, Randy Wells. Everyone was encouraging, but the logistics were up to us to resolve, and the really big hurdle was money.

Richard and I had both applied for student loans to pay for college expenses that term, but the loans had been long delayed in the processing, and checks weren't distributed until nearly the end of the term. By then, of course, we had managed somehow to survive without the money. But we hung on to those loan payments, sold our stereos, books, clothing, and everything else we could part with. By late July of 1982 we had enough money scraped together to buy two tickets to Australia.

We had been in airplanes and transit lounges for over thirty hours by the time the plane descended into Perth, Western Australia's capital city. It felt like a lifetime. Our world had been shrunken to the dimensions of those two airplane seats, the tray tables, the tiny square-windowed view of endless blank sky, the backs of the heads of the people seated in front of us, earphones spilling out of the seat pockets, the busy stewardesses. We hadn't slept a wink and had overstuffed our faces with innumerable offerings of Qantas airline's edible entertainment. We didn't feel so good.

After collecting all our luggage, I pulled out a scrap of paper on which my mother had scribbled the phone number of some old friends from her college-student days, Bert and Barbara Main. An hour later we were comfortably ensconced in the Mains' living room, which would become our Perth waystation for years to come, and a few days later we were standing on the roadside just north of Perth, with our luggage in a heap and our thumbs out.

A decrepit old station wagon, driven by a red-faced man in high white knee socks and shorts, pulled over and we loaded in. We struggled to make conversation, tripping over his heavy Aussie accent. It turned out he worked in an opal mine up at Marble Bar, way out in the middle of the Outback. Stinking hot and dry, from all accounts. He started to tell us about the aborigines he encountered there, how they would come wandering into the mining settlement. Here he opened the glove compartment of his car, revealing a .22-caliber pistol. I don't think Richard or I had ever seen a real pistol before, much less been in a car with one. "We shoot at 'em," he declared.

That night we stayed overnight in the town of Geraldton and toured the maritime museum, featuring an exhibit of remains from numerous boats that had failed to navigate through the tricky Abrohlos Islands and Zutydorp cliffs. These treacherous waters sit just below Shark Bay, the huge embayment within which lies Monkey Mia.

The following day we were back on the road again, our thumbs out. After several short rides, a ratty old Holden station wagon pulled over. Two young men sat up front, and the back was filled to the brim with junk. "No worries," said one. "We'll just move stuff around a bit. There'll be plenty o' room." Dubious, but eager to get to Monkey Mia, we got in. My place, as it turned out, was flat on my back, crammed between a heap of dirty old mattresses, boxes of junk, and the roof of the car. We roared up the Great Northern Highway at eighty miles per hour. Up front sat Richard and the two other guys, guzzling can after can of beer and flinging the empties out the window as "Steel Maiden" pulsated from the speakers located just behind my head. I shut my eyes and tried to will myself elsewhere. After what seemed like a painful eternity, we arrived at the Overlander.

Plunked down at the intersection of the main road that heads on north to Carnarvon, Broome, Dampier, and Darwin and the turnoff to Shark Bay, the Overlander road station was the only sign of human civilization for a couple of hundred miles. Almost every truck or car that heads up the coast of Western Australia must stop there, for fuel, to use the toilet facilities, or to take advantage of the diner. The station is bleak, the workers bored, the air dusty, the water fetid. In typical stoic good humor, the management had declared the fly as their "Overlander mascot." The flies here can be overwhelming. They don't bite, but rather seek fluid from your eyes, nose, and mouth. Gathering in great numbers around these parts of your face, they are persistent beyond belief. If you swat at them, at best they may jump off, fly two inches from your face, and then swing back for the return. The only way to keep them in check is to pretty much constantly swat yourself in the face and swipe at the air all around your head: the "Aussie salute." It doesn't take long before one either gets used to them or goes insane.

But I still feel a certain fondness for the Overlander. It is an essential stop along the way to Shark Bay, and each time I have made the journey, I've emerged from whatever vehicle brought me that far and been met by the familiar smells and sounds and sights that mark a return to Shark Bay country. Adrenaline courses through my veins, and I am antsy to get out to Monkey Mia.

On this first journey, I unfolded myself out of the back of the station wagon and onto the red dirt of the parking lot, grateful to be alive and with no clear idea of what still lay ahead before we reached our destination.

Just out of Perth, we had traveled first through rolling green countryside with flocks of screaming black cockatoos, vineyards, horse farms, and tall eucalyptus groves. As we headed north, the bush transformed. Now we were surrounded by short, pale sage-green scrub and brilliant rust-red dirt. Crows rattled and cawed, and a huge wedge-tailed eagle flapped down to the roadside to grab a rabbit carcass. Kangaroo remains, in various states of decay, lined the roadsides at a rate of about one for every fifty feet of highway. The kangaroos are such a hazard that most vehicles traveling outside of city limits are equipped with a "roo bar," to protect the front of the car from damage should a roo be hit, as is practically inevitable.

Other than the Overlander service station and the two-lane road, there was nothing but a vast, pretty much flat expanse of wild, desert brushland. By the time we had used the facilities and refreshed ourselves with cold drinks, the sun was low and the huge sky dotted with clouds, the undersides of which were now turning pink and orange in the fading light. We caught a ride with an alcohol-soaked roadworker who was camping ten kilometers down the road toward Shark Bay. Disgorging ourselves and all our gear from his dusty old rattletrap, we watched the taillights disappear as he pulled away and took stock of our situation. We were really out in the middle of nowhere. The fact that we had no food or water with us suddenly loomed ominously. Unless another ride came along, we were going to have to sleep on the ground out here with who-knew-what slithering and bouncing around us. I didn't have a sleeping bag, and the desert nighttime chill was already descending.

We hunkered down as best we could, and I spent much of that night shivering and watching the stars slowly roll past overhead, listening to strange, unidentifiable thumping sounds, which I later realized were kangaroos. Sometime during the night I dreamed that a strange little marsupial creature crawled in alongside of me, seeking warmth against the small of my back. When I woke up I wasn't entirely sure whether it had been a dream or not.

The morning dawned with a surreal clarity. A small group of galahs, the flashy pink-and-gray parrots, foraged nearby in a patch of wildflowers, various sizes and shapes of ants scurried and marched around and over us and our gear, and unfamiliar birds sang and hopped around in the bush, revealing themselves as flashes of iridescent blues and yellows. In the morning light the sage green brush contrasted even more dramatically with the iron red dirt, all dotted with yellow flowering acacia bushes. The landscape was so beautifully strange, I felt as if I had woken up on another planet altogether. It was a day later when we finally came over the last rise and were met with our first view of Monkey Mia, a tiny beachfront campground sitting on a northeast-facing pucker on the inland coast of Peron Peninsula, surrounded by the blue-and-green waters of Red Cliff Bay.

When we first arrived, Monkey Mia was a remote fishing camp scratched into the dirt and sand, with few amenities and no luxuries. A noisy generator ran all day, providing electricity to about fifty powered campsites: squares of dirt with outlet posts into which people could plug their RVs. Farther along were cheaper, unpowered campsites for backpackers. A temperamental pay phone hung in a booth under the radio antenna at one end of the grounds. A spare brick building, the "ablution block," housed toilets and showers featuring salty water.

Denham, a sleepy little fishing village eighteen miles away on the other side of the peninsula, was the closest town, but with no car even that seemed remote. Denham had two shops, one of which doubled as post office. Mail was flown in twice per week to Denham International Airport, a short strip of dirt and a corrugated tin shed. Denham also featured a pub and church and a small golf course carved into the red dirt.

Leaving our gear in a heap, we located Wilf and Hazel Mason, the proprietors of the camp, whom we had corresponded about our visit. Wilf and Hazel had moved to Western Australia from Queensland in 1975. As permanent residents, they had fallen in love with the dolphins and had begun to recognize the potential tourist attraction that could be developed. Wilf, in his early sixties, sported a protruding beer belly, a bright red sun-tortured complexion, small gleaming eyes, and pure white hair. He worked incredibly hard, long hours to keep the camp, such as it was, up and running. Hazel, a small woman with huge brown eyes set over a mouth prone to scowling, tended the small shop where visitors could purchase frozen milk and bread, a can of beans, or a chocolate bar.

Monkey Mia was inhabited by a small group of "regulars," retired folks (we referred to them as "the oldies") who had been coming to Monkey Mia for a few weeks or months each year for as long as anyone could remember to fill their freezers with fish and enjoy Shark Bay's charms. They were all friends with a history of sharing the camp and looking out for one another. They were also a somewhat insular bunch who looked upon newcomers such as ourselves and the expanding crowd of tourists with some disapproval. They had grown accustomed to having Monkey Mia to themselves, and in their view, the attraction was fish and peace. The brouhaha over friendly dolphins was difficult to comprehend.

Our correspondences with the Masons had been full of warmth and welcoming, so it was with some sense of disappointment that we were abruptly escorted to our campsite, at the farthest end of the camp-ground, where we pitched our tents on the sand, alongside a scrubby lit-tle acacia bush, and dug out a fire pit in which to cook our evening meal ("tea"). With the journey behind us and the dolphins ahead, we fell into our tents and were introduced to the Shark Bay wind as the tents rattled and flapped and all but sailed off down the beach but for the weight of our bodies inside.

I did my best to sleep but before dawn was awakened by seagulls screeching at each other over contested morsels of fish. Richard and I stumbled out of our tents and down to the water's edge. A dog lay in the sand, staring out at the water, and a couple wrapped up together in a blanket sipped steaming cups of coffee. We waited, watching the sky turn from deep indigo blue to pale pink to brilliant orange as the huge globe of the sun rose and dislodged itself from the watery horizon. Then *phoohoof*, the sound of dolphins breathing just twenty yards from shore. The dog, whose name we later learned was Ringer, got up and raced into the water. We followed, yanking up our pants legs, kicking off our sandals in the sand. The chill of the water woke us up more effectively than even the strongest cup of coffee.

The dolphins turned toward us and approached. The larger dolphin, with a small hole in her dorsal fin, swam directly toward us, stopped a foot away, and lifted her face out, looking us each in turn directly in the eye. Big, dark, smart eyes set in a rubbery gray face. We knew from Elizabeth Gawain's slides that this must be Holeyfin. She opened her mouth, revealing neat rows of small pointy teeth, and continued to look us over expectantly. Then she glanced back at the other dolphin trailing behind, her two-year-old daughter, Joy.

I reached out slowly and tentatively and touched her side. She watched me intently but did not flinch or move away. I was stroking the side of a wild dolphin. Her skin was silky smooth, slightly rubbery, and surprisingly warm for a creature living in the ocean and resembling a fish. I felt suddenly aware of how odd my long, gangly arms, with all those independently moving digits, must seem to her. She was sleek, a torpedo. No wiggly, flapping, dangling, wrinkly, stretchy, or sagging bits. Just a fused cylinder. A head tapering down to a tail with a few simple fins attached. Streamlined and elegant. After a few minutes, she turned and with one effortless sweep of her tail glided off into deeper water, toward Joy.

FIRST

VISIT

Our first days were spent trying to learn how to identify the dolphins that came into the shallows. At the time, there were seven who visited Monkey Mia on a daily basis. All could be distinguished by natural markings, nicks, and scars on their dorsal fins. They had been given names by Wilf and Hazel, who had kept some records of births, deaths, and other events. At first it was all I could do to discriminate among the dorsal fins of all the different dolphins, but quickly they became second nature and I learned to recognize their faces, bodies, and mannerisms as well. Over the years we would come to know the personalities of these dolphins, each as distinctive and quirky as any human being.

HOLEYFIN

Holeyfin, named after the pea-size hole through the middle of her fin (some claimed from a bullet), was the grand old matriarch of Monkey Mia. In 1982 she was accompanied by her daughters, Joy, two years old and still dependent on her mother, and Nicky, seven years old. When we first arrived, she already seemed to be an old woman with worn-down teeth, fishy-smelling breath, and a slightly daffy, senile demeanor. She had given birth to Nicky in 1975, and since female dolphins generally give birth to their first offspring at around twelve to fifteen years of age, she had to be at least seventeen to twenty years old during our first visit. If Nicky wasn't her first, she may have been considerably older.

Perhaps the most prominent feature of Holeyfin's behavior was her dogged pursuit of fish handouts. She made a point of approaching each and every person who waded into the Monkey Mia shallows. Mouth agape, worn-out teeth in full view, she was perpetually begging for handouts. When that failed, she would follow each and every boat that came or went from the mooring area, swimming alongside, lunging upward to put her face out, mouth open, at the passengers aboard, even as boat and dolphin were under way. Her persistence sometimes seemed touchingly pathetic, at other times wily and dogged. I was always surprised to see her away from the Monkey Mia shallows, hunting for herself in the company of the wild dolphins. She was a hard-core Monkey Mia dolphin, probably the most photographed, touched, and well-known (by humans) wild dolphin on the face of the earth. It was sometimes easy to forget that she was actually a wild animal.

NICKY

Nicky, Holeyfin's daughter, was an adolescent when we first arrived at Monkey Mia. A big, stocky dolphin, named for the several sharp nicks out of the top of her fin, she impressed me as being something of a tomboy dolphin. She seemed to relish the rough-and-tumble antics of the local gang of young males: Snubnose, Bibi, Sicklefin, Wave, Shave Lucky, Pointer, and Lodent. Temperamental and moody, she often seemed irritated by the attentions of visiting tourists. Although she was more than willing to accept fish handouts, she wanted nothing further to do with us much of the time. When a hapless tourist tried to stroke her side, she was likely to toss her head and snap at a hand. She drew blood on more than a few occasions. But then she would have one of her moods: gushing with affection and fellowship, she would choose some lucky sod to bestow her affections upon and then lay it on. One such incident involved Richard. She sprawled languorously across his lap, wrapped in an embrace, eyes half-shut, enraptured. She nibbled gently at his toes, rolled belly up to permit him to stroke her, allowed fingers to explore her chin, around her blowhole, under her pectoral fins. She was on. There was no telling what determined her moods, but one thing for sure: she was unpredictable, smart, and complicated. She was also an avid observer of human behavior.

CROOKEDFIN

Crookedfin, another adult female, named for her bent-over fin, brought in her six-year-old daughter, Puck. She was always a bit shy. It took some patience to get a fish handout to her, and she seemed a bit unsure of how to ask for one properly (the more experienced dolphins would approach a person, brace themselves against the sandy bottom with their pectoral fins, arch their backs to raise their heads up, and open their mouths: the begging posture). Inevitably, when we encountered Crookedfin offshore in our boat, she would approach us and release a big bubble of air from her blowhole, a sign of nervousness. I wondered if she had some bad experience with people that left her feeling a bit ambivalent. When we first arrived at Monkey Mia, Crookedfin came to the beach every day, but her visits became less and less frequent over time.

PUCK

Unlike her mother, Puck was far from shy. She was graceful and sleek, with a long, delicate beak, soft eyes, and perfect skin: an exceptionally beautiful dolphin. She was also the sweetest, most even-tempered of them all, unlikely to bite or otherwise be aggressive toward people. That is not to say never; even Puck had her limits. A vivacious adolescent when we first arrived on the scene, she and Nicky, being close in age, were good friends, and together they spent much of their time playing with "the boys." The two young females and their entourage of boyfriends spent hours just offshore of Monkey Mia, splashing, chasing, flirting, and roughhousing. Abruptly, Nicky and Puck would break away from the males and head in to the shallows, where they would eat a few fish and entertain the humans for a bit, while the males waited impatiently for them to come back out and resume their games.

BIBI

Another adult female, Beautiful, had disappeared the year before our first visit, leaving her adolescent son, Bibi (short for Beautiful's Baby).

To Touch a Wild Dolphin

Bibi was easily excited, often aggressive, a bit shifty, and unpredictable. His small, bright eyes seemed to barely conceal a touch of madness, accentuated by a slightly twisted rostrum tip. I often had the feeling that he was of rather low status and that knowing this drove him crazy. For example, one day I watched as he was interacting with people and growing more and more agitated. Then he approached and, facing me, began to emit a wild assortment of squeaky, squealing, splatting sounds while bouncing himself up and down on his erect penis against the sandy bottom. He was always hard to figure.

SNUBNOSE

Snubnose, a mature male, also was coming to Monkey Mia when we first arrived, but nothing was known about his familial ties. Interacting with people seemed to be second nature for all of the other dolphins except for Snubnose. He was uneasy and hung back, keeping his distance from people.

With some careful coaxing he could be enticed to take a fish from the hand, but this required wading out into deeper water to get close to him and gently, quietly holding a fish in the water just in front of him. Sometimes he would take it from our hands, but more often than not it was necessary to release the fish or throw it out toward him a little. Slowly but surely he became more and more accustomed to people, and eventually he became quite at ease. Snubnose reminded me of the sort of human male who grows a bushy beard and pot belly, always has a joke ready for the telling, and loves to play with the children. Everyone's favorite uncle.

With his comically upturned lower rostrum tip, from whence he got his name, and his big brown eyes, Snubnose was inclined to be gentle, calm, trustworthy, and a bit goofy.

This was the "cast of characters" at Monkey Mia when we first arrived. For the most part, the Monkey Mia dolphins were mothers (Holeyfin, Crookedfin, and the recently disappeared Beautiful) and offspring of those mothers who grew up accustomed to visiting with people at Monkey Mia. There were only two dolphins without family histories of visiting Monkey Mia who, in later years, did become regular visitors. One, Sicklefin, was a male with a strong bond to Snubnose and Bibi. The second, Surprise, was, from the beginning, particularly friendly toward people in boats and made occasional visits into the Monkey Mia shallows, usually in the company of one of the "regulars." Eventually she too became a regular.

Dolphins came and went from the Monkey Mia shallows all day long. When they were in, Richard and I waded into the shallow water to play with them, to watch them, and to record everything we observed on our little handheld tape recorders, which we would eventually transcribe into notebooks. The dolphins allowed us to touch them, but they were particular about where and how. Not on the head, especially around the blowhole or eyes, and not on the chin or belly unless specifically solicited (by rolling upside down to expose these areas). The dorsal fin and pectoral fins also seemed to be off-limits. They let us know their preferences by tossing their heads, snapping at hands, karate chopping against shinbones with their pectoral fins, or, more mercifully, moving out of reach.

Although simply feeding the dolphins by hand was certainly a thrill at first, I wanted to engage them in some way that would provide greater opportunities for fun and games and hopefully some insights into their minds. Seagrass games fit the bill. Long flat strands, which break off and wash up onto the beaches in great piles, provided an endless supply of toys. The dolphins' games, played with us and with one another, consisted of endless variations on a theme of "give and take," like the following:

Nicky and Puck were in the shallows. Richard and I, along with a couple of kids, were standing in the water. When Nicky came by me, I looked around for a piece of seagrass to play with. The closest one was only about an inch long. I picked it up and offered to her the teensy-weensy little end. She approached and very gently and deftly took hold of it in the tip of her rostrum. When I didn't let go immediately, she gave it a small tug, then turned an eye to look at me as if to ask "What's the deal here?" I tugged back, she tugged back, I tugged, she tugged, and this time I let her take it. She immediately dropped it and moved on. The act of taking it from me was the game.

Another time I found myself surrounded by strands of seagrass floating all over the water surface. Still, Puck wanted the very piece I offered to her, even though there were others identical to it floating right in front of her face. In fact, when I tossed the piece I had in my hand, and it landed among many other pieces, she picked out exactly the one I threw and carried it around before approaching and offering it back to me. We repeated this game seven more times, and each time she picked out exactly the piece of seagrass that I threw from among all the others. On another morning, Snubnose was having a grand time taking bits of grass from Richard. The game went like this: Snubnose feigns disinterest, approaching other people, begging for fish, and so on, but you can tell his attention is really on Richard. Then he "remembers" and makes a dash at Richard. As he gets close, he rears the front half of his body out of the water, lunging up to take the piece of grass, but Richard moves off and starts walking backward away from Snubnose, who follows him the length of the beach. Eventually he gets the grass from Richard, drops it, loses interest again, and repeats the cycle. On a couple of occasions Snubnose turned upside down, chasing Richard with his belly to the sky. Very goofy. The two of them repeated this game over and over again all morning long, with Snubnose growing more and more exuberant.

The time we spent playing with the dolphins was not time wasted. We learned to read the dolphins' signals, to see subtle aspects of their posture and demeanor that contributed tremendously to our over-all sense of them. We were honing our skills as dolphin observers. These friendly and accessible dolphins were providing us with a window into their world through which we could see things that would be impossi-ble to see out in the bay, in deeper water, with dolphins less accustomed to humans.

How and when dolphins and humans first made friends at Monkey Mia remains a mystery, but certainly it started long before we ar-rived. Some of the old-timers told stories about the dolphins who used to visit in the early days. Specifically, they remembered Old Charley, a dolphin who apparently used to come into the shallows during the 1950s. As the story goes, Old Charley would herd schools of herring into the shallows so that the fishermen could catch them easily, and as a re-ward for his effort, the fishermen would give him a few of the fish they caught. There were reportedly other dolphins that came in to the shal-lows at that time, but Charley was the best known. I have my doubts about generous Old Charley. Even now the dolphins sometimes chase fish into the shallows at Monkey Mia, in proximity to people. Those who have witnessed this often claim that "the dolphin chased the fish in to me," or "the dolphin gave me the fish." I've witnessed this on a number of occasions, and to me it looks as though the exhausted and disori-ented fish is seeking refuge in shallow water and is pursued there by the dolphin. There are some well-documented cases of wild dolphins hunting cooperatively with people (in Brazil, Mauritania, and along the east coast of Australia, for example). In Shark Bay, I suspect, the dolphins were not necessarily cooperating with people but just happened to be in proximity of the people who took advantage of the situation.

In the early to middle 1960s there was a dolphin called Speckledybelly, a "toothless old woman" of a dolphin who hung around the shallows at Monkey Mia. This dolphin was supposedly very friendly with people. After her, other dolphins carried on the tradition.

An elderly woman by the name of Nin Watts claimed to be the first person to feed a dolphin by hand at Monkey Mia. She had stayed for some time on a sailboat moored just offshore from the campground in the mid-1960s and coaxed a dolphin to take a fish from her hand. I'm just not sure how she could know that she was the first, that nobody else had fed a dolphin at Monkey Mia before her, but that's what she claims.

I view all of these old accounts with skepticism, having seen how easy it is for dolphin identities to be confused and for stories or impressions to turn into "truths." Just as an example, I have watched countless visitors, having heard rumors that the dolphins could be distinguished by nicks on their dorsal fins, then assume that any dolphin with a nick was Nicky. And one old fisherman, a wizard with respect to fishing and fixing just about anything mechanical, used to tell us about seeing Holeyfin and Nicky at various locations all over Shark Bay, way beyond where those dolphins range, and at times when we knew they had been right here in the shallows at Monkey Mia. Somehow it hadn't occurred to him that there were many other dolphins in the bay. Anytime he saw a dolphin, no matter where, he assumed it had to be one of the Monkey Mia dolphins.

Without careful study and a concept of the range of variation in dorsal fins, as well as some background knowledge about dolphin behavior, people arrive at some wild misconceptions, which are then handed down eventually to become "historical fact." It is easy to imagine that once people heard from others that there was a friendly dolphin at Monkey Mia called Old Charley, they would assume any friendly dolphin they encountered at Monkey Mia was Old Charley. Without photographs of his dorsal fin to compare with Speckledybelly, the possibility remains that the two were one and the same or many different dolphins.

The true origins of human-dolphin interaction at Monkey Mia are unknowable at this point. Maybe the aborigines started it, or maybe old Nin Watts really was the first to feed a dolphin at Monkey Mia. Maybe Old Charley was the original Monkey Mia dolphin, or maybe he was just the one who was around at that point in time which is now the limits of memory for the oldest of the old fishermen from whom we heard stories. With all the different stories and plausible scenarios, we can only guess that all may have contributed some part to building the relationship. My own guess is that the dolphin-human relationship at Monkey Mia was started by the local fishermen. Monkey Mia is within Shark Bay and is one of the only places where relatively deep water abuts the shore. The local fishermen have long taken advantage of this fact to bring their boats up to shore to clean out their nets. Just as they do now, they have probably always thrown a few fish to the dolphins while they were at it.

Interacting with people was apparently an acquired skill. An experienced dolphin approaches a person who has a bucket of fish and assumes the specific begging posture: pectoral fins braced against the bottom, head held up and out of the water, mouth open. Inexperienced dolphins seemed to find this awkward and would sometimes "practice." Joysfriend, a dolphin who became very familiar to us from our offshore observations, and who spent a lot of time with Holeyfin's daughter, Joy, did not usually come into the Monkey Mia shallows. But occasionally she visited with Joy and Nicky and Puck. Once I watched her swim into shallow water at the shoreline, facing an empty beach. Without putting her face up, she stopped and looked around expectantly, as if to say, "Okay, so where's the fish?" She moved back offshore and seemed to watch the other dolphins approach people and take fish. A while later she tried again, this time bringing her face up out of the water but failing to open her jaws. She was going through some of the motions, but again, she was several yards from the nearest person and still didn't have her jaw open. She looked awkward, like a person taking his or her first stab at stage acting.

The habituated dolphins sometimes came into Monkey Mia accompanied by companions from offshore who normally did not come in. With a few interesting exceptions, these newcomers would usually hang back a bit. They weren't accustomed to contact with people and generally did not take fish handouts or permit physical contact. These rare visitors took an interest in what was going on at Monkey Mia, but they seemed to be at odds, both curious and shy. Perhaps they had "heard about" the strange beings who live on land but wade into shallow water to bestow free handouts on a few privileged members of their community.

I once had the opportunity to feel what it is like to be an animal curiosity. Several dolphins, a mix of both Monkey Mia regulars and a few who were unfamiliar to me from offshore, were chasing fish in shallow water just offshore of the far west end of the campground. I waded out to watch them more closely and found myself standing in the middle of a feeding frenzy. A large school of fish swam around frenetically in the shallow water, and dolphins were tearing around me on all sides. Silver gulls and Caspian and Royal terns screeched and reeled overhead, diving down on the fish from above. In the shallow water, I could see the fish clearly and occasionally even caught a glimpse of a dolphin chasing and snapping up a fish in its jaws right alongside of me. After half an hour, the fish school dissipated and the frenzy died down.

Snubnose, who had been busily hunting up to this point, approached me now and put his face out of the water by my hand. I stroked him and spoke friendly gibberish. Then he put his head back under the water, turned to face offshore toward some other dolphins, and whistled very loudly. Two of these other dolphins, both of whom were strangers to me, were about forty yards out but turned at the whistle and came directly toward Snubnose and me. They came up alongside Snubnose, placing him between themselves and me. The two strangers seemed nervous. They were "holding hands"—that is, they had their pectoral fins held against each other's sides—and their movements were fast and a bit jerky. The three dolphins circled me four or five times, eyeing me intently the entire time. Then they swam off, leaving me with Snubnose. I felt like an exhibit in a zoo. These dolphins, apparently on Snubnose's invitation, had come over to have a close look at the human. And once they had taken a look, they swam off again.

Did I make an okay impression? Had they ever seen a human being before? What might they think about their companions, Snubnose and the other habituated dolphins? Did they think them brave for being unafraid of these aliens? Or dumb for allowing the aliens to touch them? Did they wish that they too could get fish handouts? Or did they view the Monkey Mia dolphins as social pariahs, lowering themselves to eating dead fish rather than catching their own? Did the Monkey Mia dolphins somehow keep others away from the campground area, defending it as their territory? Or was Monkey Mia considered the dolphin equivalent of Siberia?

Richard and I longed to get offshore on a boat and get to know some of the other dolphins in the bay, those who did not come to the Monkey Mia shallows. But we didn't have a boat and so were limited to watching from shore with one exception: Wilf and Hazel allowed us to make use of their small dinghy for two September days. Those offshore excursions proved to be incredibly valuable. It was abundantly clear that Holeyfin, Nicky, Joy, Crookedfin, Puck, Snubnose, and Bibi were part of a much larger dolphin population. We encountered group after group of dolphins. Some even approached us to ride at the bow, looking up at us from beneath the surface and hanging around close to our boat, apparently unafraid of people. We were easily able to photograph their dorsal fins for identification purposes and, on later sorting through the pictures, realized that we had seen close to one hundred different dolphins. This was our first introduction to the "offshore" dolphins, many of whom we would get to know well and whose lives we would follow for years to come.

We decided early on to keep records of how much time each of the dolphins spent visiting the campground. There was no way we could watch the beach all of the time and keep track of all the comings and goings of each dolphin. Instead we would sample systematically. Every half hour we would jot down which dolphins were in the shallows and note any dolphins we could see just offshore. We carried tape recorders in our pockets, and whenever something interesting happened, we described it blow by blow into the recorder. At first we tried to watch everything and to describe it all in great detail. Over the years we would hone our skills as observers and learn how to focus on the kind of information that would really let us answer some questions. Nonetheless, even on our first visit to Monkey Mia, we made some intriguing discoveries.

Puck was about four and a half to five years old at the time of our first visit. We knew this because Wilf and Hazel Mason had recorded the date when Crookedfin first appeared with a new baby. So it came as quite a surprise when we saw her nursing from her mother in the Monkey Mia shallows. Virtually all of the preexisting data on age of weaning in dolphins suggested that mothers usually weaned their youngsters by about eighteen months. Puck was way out of bounds! Even Joy, much younger than Puck at two and a half, should have been weaned by now according to the current literature but was still nursing from Holeyfin.

The standard wisdom regarding age of weaning for bottlenose dolphins—namely that it occurred within the first year or two—was derived almost entirely from a few observations of captive dolphins in oceanaria. Over the years we came to realize that it is typical for dolphins in Shark Bay to continue nursing for about four years. One youngster we watched continued to nurse for at least six.

Even after four years of nursing, a dolphin is still just a baby. Females don't reach sexual maturity (the age at which they may become pregnant for the first time) until they are about twelve years old. It's harder to tell for males, but they probably don't reach sexual maturity until even later. We had no idea of birthdates for most of the dolphins (with the exception of babies born since we began watching in 1982 and those whose birthdates Wilf and Hazel Mason had recorded). Determining ages therefore required some "guesstimating," based in part on the dolphin's size and also on the extent of its belly speckling. Dolphins in Shark Bay are born with pure white bellies. Conveniently for us, around the time they reach sexual maturity, they begin to develop a few gray speckles on their bellies. By the time they are fully mature, middle-aged dolphins, their bellies are covered with speckles, even extending up their sides. The maximum life span for females is probably somewhere around fifty years and for males probably a bit less. The time frame of a dolphin life is therefore not too different from that of a human. We are both long-lived creatures with a large portion of our life span taken up by the juvenile stage.

How much time an animal spends growing up depends on many factors, but one thing is certain: In evolutionary terms, all else being equal, it pays to grow up as fast as possible. The longer one dawdles, the more risk one is exposed to prior to reaching sexual maturity, the higher the possibility that one will not survive long enough to reproduce.

But of course there are many other considerations in determining the course of an animal's maturation: how long it takes to reach adult body size given certain ecological constraints, what the animal needs to learn in order to be a successful adult, and even seasonal constraints on growth and reproduction. Dolphins and people have remarkably long life spans. Long-lived animals tend to have long juvenile stages and (in birds and mammals) big brains. The two are related, since it takes time to teach a big brain how to be a proper adult. Baby dolphins must have a lot to learn.

One October afternoon during our first visit to Monkey Mia, as Richard and I were watching dolphins come and go from the shallows, Snubnose suddenly took off at top speed parallel to the beach. We trotted after him as he stayed just a little bit offshore, apparently in hot pursuit of some fish. Then, to our amazement, a great long fish leaped out of the water and skittered along on its tail, "tailwalking," just in front of Snubnose. He tried to snap at it, but the fish escaped after tailwalking a distance of ten or fifteen feet. Again Snubnose accelerated, and the fish evaded. After a few tries, Snubnose did manage to grab the fish, and we were able to get close enough to see it more clearly. A big Long Tom. Long Toms are a type of garfish. Long and thin, this particular species grows up to a yard or so in length. They have great bony heads crammed full of needle-sharp teeth. Besides the teeth, their blue-green bodies are coated in a thin slime that smells awful. The fishermen brought these in from their nets once in a while and gave them to us to feed to the dolphins. Untangling them from a fishing net, as I later discovered, was no fun, and the stink from handling them lingered in spite of soap and water.

Snubnose's success immediately attracted the attention of Bibi and the other dolphins. They approached him closely, as if admiring the fish, faces close to Snubnose's as he held Long Tom crosswise in his jaw. Then Snubnose dove, and we could see from the way his back humped up that he was rubbing the fish back and forth against the bottom. When he resurfaced, the other dolphins were still gathered close around him. Were they hoping to get a piece? Snubnose threw the fish with a snapping motion, and we could see that the fish's head was dangling now. After another dive, closely attended by the other dolphins, Snubnose surfaced without the fish. He'd eaten it, perhaps with a little help from his friends.

We began to wonder whether the dolphins might share these big fish around. Chimpanzees, especially after hunting and killing a source of meat (like a monkey), gather together in a tight little cluster around the captor and "beg" for a share, holding a beseeching hand out toward him or pressing close to his face. The dolphins' behavior reminded us of begging chimpanzees and caught our attention.

Sharing food may seem commonplace to us, since we do it often and barely give it a thought, at least when there is plenty to go around. But sharing one's food is costly, especially when there is not a lot to spare. Genetically speaking, it generally doesn't make sense to incur those costs, but there are a few scenarios in which food sharing could evolve. First, if the animals you are sharing with are closely related to you, the behavior of sharing could evolve because you have genes in common with your kin. The cost of sharing is lessened by your relatedness to them. You (your genes) are doing yourself (the same genes in the body of your relative) a favor. So mothers and fathers feed their offspring, and we can expect that brothers and sisters are more likely to share food than are unrelated individuals. But some of the dolphins involved in these garfish hunts were unlikely to be close relatives.

Another situation where food sharing might evolve is when you can't get the food unless you share it. Take the hyena, for example. These animals are hunting machines with jaws that easily crush the bones of a zebra. But a single hyena cannot take down a large prey animal on its own. Hunting together enables them to take larger prey than they would otherwise be able to catch. Although they may have to share the kill with unrelated individuals, they still come out ahead. But the dolphins were not doing this. The actual capture of the garfish was an individual achievement.

One other situation could result in the evolution of food sharing: reciprocity. Reciprocity is a slippery concept and notoriously difficult to recognize or prove. The basic idea, however, makes simple, intuitive sense: I do something for you now, with the understanding that you owe me a return favor later. We don't have to be relatives or "co-dependent." What we do need is to be able to recognize each other and be confident that we will run into each other later. Of course, it helps if there is some way to enforce the return favor.

Although we know from experience that humans engage in reciprocity, it has never been demonstrated conclusively in any other animal. We expect it to happen only in social, group-living animals where individuals interact with each other repeatedly. Dolphins fit the bill in that sense. If they are sharing, can they be doing so with the expectation that later they will receive the same courtesy (or one of similar value)? We couldn't be sure, but these observations of Long Tom hunting suggested to us that dolphin social relationships might include some interesting "fish politics," and we were on the alert for further observations.

Meanwhile, we were not the only people interested in the dolphins of Monkey Mia. As word got out, more and more people rose to the challenge of getting out to Shark Bay to see the dolphins. With increasing numbers of visitors, tourism developers had dollar signs in their eyes while concerns were mounting about potential threats to the welfare of the dolphins.

Tensions mounted as we began to voice our dissent with respect to some of the dolphin-human interaction practices. In particular, we were concerned that Wilf and Hazel were selling little packets of "mulies," small, oily, anchovylike fish, to the tourists to feed to the dolphins. The fish were frozen and were intended to be used as fishing bait. They often showed signs of decomposition and, besides, were not a species that occurred naturally in Shark Bay. The dolphins gobbled them down, even half-frozen. In fact, though they tended to turn up their rostrums at butterfish (fresh caught and local), they seemed otherwise pretty indiscriminate about what they would eat.

Once, a woman on the beach asked me what dolphins liked to eat. I explained that they preferred fish and probably also ate some crustaceans and other marine organisms. She pulled out a shrimp salad sandwich, replete with mayonnaise, and began dropping bits of the bread and salad onto the water surface, as if to attract fish. I explained that this might not work, that they only ate fresh, raw seafood, and that she would have better success if she waited until a dolphin was actually present. A little later I came down to the beach to find Nicky in. She swam up to me and regurgitated two cooked shrimp along with a cloud of guck.

We worried that the dolphins would get sick. In their natural world, they would probably never eat anything but the freshest fish—still alive, in fact, and certainly not half-frozen or decaying. People argued that the dolphins were smart enough to know better than to eat anything that might not be good for them. But it seemed quite likely to us that since they would not normally be faced with frozen or rotten food in the wild, they might have no evolved capacity to distinguish rotten from fresh food.

The sale of these packets of bait fish was turning into a major source of income as more and more tourists arrived, seeking to have their picture taken feeding the dolphins. We suspected that this financial concern might be at the bottom of the resistance we faced in trying to change dolphin-feeding policies: the dolphins were being stuffed full of junk food.

We also worried that someone might hurt the dolphins accidentally or even intentionally. In other places where people and dolphins had taken to interacting, the dolphins had often enough ended up dead. One such case occurred in Opononi, a tiny, sleepy fishing community in New Zealand, where a young child had befriended a dolphin, word got out, tourists began flocking to town, and not long afterward the dolphin was found dead. Word about the Monkey Mia dolphins was already spreading far and wide. The road into Shark Bay would soon be paved all the way from the Overlander to Monkey Mia. Many more people would be able to make the journey. This isolated, lazy little place was poised on the edge of radical changes, and there were many conflicting opinions about how to progress.

One thing everyone (Wilf and Hazel, the local government, the tourism interests, and ourselves) agreed upon was the need for some presence on the beach to monitor human-dolphin interactions. Someone needed to make sure that nobody did anything to harm the dolphins and also to field questions that visitors might have about the dolphins.

A small trailer was donated by the "Golden Dolphin Tribe," a band of hippies who had made a pilgrimage from Sydney across Australia to visit the "mystical" Monkey Mia dolphins. The trailer was marvelously decorated with a painting of dolphins leaping in front of Ayers Rock (which lies dead smack in the center of Australia: as far from ocean as one can get on the continent). Richard and I were encouraged to avail ourselves to the tourists to answer questions about dolphins, to help keep an eye on things, and to aid in developing the trailer as an interpretive center.

One day we watched as a small dog made its way to the water's edge, looking intently at the dolphins as if planning an attack. It then plunged into the water and swam out after Puck. Puck had no trouble at all staying out of the dog's way. She glided ahead of the dog and seemed to be teasing it by staying just out of reach as the dog furiously paddled, huffed, and snorted, head barely above the surface. After leading the dog farther and farther offshore, Puck then rushed back into the shallows, leaving the little dog to struggle back to shore alone. Though it was clear that Puck could handle the situation herself, Richard and I felt obliged to enforce the newly agreed-upon rule that dogs were not to go into the water near dolphins.

So we made our way down the beach to an elderly man who was obviously the dog's owner. He wore a stained and tattered T-shirt and shorts and stood with thin, hairy legs held slightly apart, planted firmly in the sand. His face, covered with ragged gray hair, sported two reddened and beady eyes that peered suspiciously from under the brim of a captain's hat as we approached. His demeanor said, "I dare you." But Richard and I were too naive to notice. Meekly I requested, "Could you please keep your dog out of the water and away from the dolphins." His name, we later learned, was Bill Bond, known to many as "Bondi," and he let loose with the nastiest bunch of cursing and berating I had ever experienced. "You f—— Yanks think you can come in here and tell me what to do. . . . I'll do whatever I f—— please. You come around telling me what to f—— do and I'll put corks in their blowholes, I will . . . ya little s—heads . . ." And so on. Saliva flew out of his mouth, and his face turned from red to purple. The dog exited the water, shook himself off, and trotted close at Bondi's heels as he made his way back toward their campsite, still fuming and cursing.

Bondi was one of the Monkey Mia oldies and had been coming to Monkey Mia for much of his life. Needless to say, he did not take kindly to having a couple of newcomers—foreigners, even—telling him what to do. Bondi set himself against us with all his mighty force, and from that day on we tried hard to avoid crossing his path. When we did, he looked askance at us, muttering all variety of profanity. He had a lot of friends among the oldies, and trouble with Bondi could only spread among the rest of the community.

In June of 1982, having braved the shallow and poorly charted waters of the inner part of Shark Bay, a yacht rounded Cape Rose, entered Red Cliff Bay, and finally pulled up to the jetty at Monkey Mia. Her name was *Aloha*, a weather-beaten old schooner-type rig with her hull painted a dull and chipping blue. The crew consisted of an older couple, Graham and Jackie, not of the new breed of wealthy, sophisticated, high-tech sailors, but a couple of true "salty dogs." Graham was a small, wiry man with a scraggly gray beard and bushy eyebrows. Jackie, a small woman with a thick waist and thin legs, had hair dyed black except where it had grown out gray at the roots and toenails painted bright red.

Richard and I quickly became acquainted with Graham and Jackie on the beach and over the following weeks enjoyed sharing meals and going sailing with them. One evening we got to talking about why it was that dolphins love to bowride on boats. Were they "hitching" a free ride? That seemed unlikely, since they often seemed to go out of their way in order to ride and then had to backtrack afterward. Was it purely for fun? I suggested in jest that we should try it sometime, that maybe the experience of doing it would provide some hints. The next day Graham had a proposal. He would make a harness for me and lower me down off *Aloha*'s bowsprit while the dolphins were riding at the bow. I could brace my feet against *Aloha*'s hull and ride the bow with the dolphins.

Conceptually it seemed straightforward, but in practice it was far from it. The hull was slippery, and I had a hard time bracing my feet. The forward momentum of the boat meant that water streamed back against my face, pouring into my snorkel, dislodging my mask, and pushing me toward the rear of the boat. Bubbles impeded any view, and it was often difficult to simultaneously keep from drowning and clearly signal my distress to the person driving the boat. Our first few attempts were comical. After clambering awkwardly over the bowsprit, I would find myself dangling helplessly while Richard and Graham and Jackie all heaved and hoed to pull me back on board.

In retrospect, it probably wasn't the safest thing to be doing. Besides the risk of drowning, there was also *Aloha*'s propeller, slicing through the water just a short distance behind me. But when the moment of reckoning arrived, it was all worthwhile. Nicky and Puck approached the bow, and I was lowered into position. As they glided gracefully through the water alongside me, I gurgled into my snorkel and strained to keep my feet braced against *Aloha*'s hull. It was far easier for them than for me, but having gained some confidence, I stretched out both arms and pressed one hand against Puck's side and the other against Nicky's. In my wildest dreams I had never imagined doing such a thing, but here I was, bowriding with two wild dolphins!

By October, with the Southern Hemisphere's summer season approaching, temperatures were soaring. The sun blazed, and the glare off the water became painful. Richard and I were both red and dry and raw from the sun and salt. The wind blew day after day after day, and the dolphins visited less and less frequently, as was apparently their pattern at this time of year. Our little tents on the beach became unbearable ovens into which we dared not venture until well after dark. One morning I awoke to a scorpion, perched a few inches above my face on the tent ceiling, tail curled ominously over its back. I grabbed my thong and pounded it into an unrecognizable smear, in the process poking a hole right through the sun-ruined nylon of my tent. It was time to go.

This first visit to Monkey Mia had exceeded all our expectations. In just three months, from early August through October, Richard and I had made a few interesting discoveries about the dolphins' behavior, and we had experienced the great satisfaction of interacting with wild dolphins, touching them, playing with them, getting to know them as individuals. It had been a blast, and at the same time, we had learned more than enough to form impressions and develop ideas for further research. It was abundantly clear that we would be able to collect systematic data on the dolphins' behavior here in the shallows at Monkey Mia. We were also convinced that if we had access to boats, a world of opportunities and discoveries lay offshore as well. We had barely scratched the surface, and potential seemed virtually infinite. We left full of ideas and determined to return.

RETURN

TO THE

DOLPHINS

A fter our first visit to Monkey Mia, I spent several months visiting other points of interest in the Southern Hemisphere and then arrived back in Santa Cruz, befuddled, disoriented, completely broke, and with no place to live. Richard and I had amassed several binders full of notes about dolphin behavior during our first visit, and now our task was to make some sense of them, write up a report, and begin applying for research grants that would allow us to go back.

But first I needed to get my life in order. I was begging couch space at the homes of friends and acquaintances, carrying a few essential personal items around in a small suitcase. But more than that, I had seen things and done things that I couldn't really share with my friends. I had changed. My absence had spelled the demise of a long-standing romantic relationship. His life and those of my other friends had gone on without me. Now that I was back, we spent time together, but my obsession with going away again, back to Australia, left us with guarded affections. I just didn't fit into my old niche in Santa Cruz. My roots were pulling up, and I was miserable and lonely.

Finding a place to live was imperative. Otherwise my sense of being a stranger in a strange land would only grow and fester and interfere with my plans. I borrowed some money from my family and bought a 1959 VW minibus badly in need of paint. The most distinguishing feature of this bus was not its age—there were quite a few old VW buses around in Santa Cruz—but its red Naugahyde tuck-and-roll upholstered interior. It was almost a work of art—studded and cushioned and fantastically tacky. To complete the piece, I bought some peppermint green paint (it was cheap) and spray-painted the exterior. The combination was stunning, an enormous synthetic watermelon on wheels and definitely one of a kind.

That bus became my portable, rent-free home, and in some ways, my future ability to go back and forth to Australia depended on her. When I left for Australia, I would pack my worldly possessions into the bus and park it somewhere safe. When I came back to the States, my rent-free home awaited me. Ken Norris generously provided me with office space at UCSC's Long Marine Lab, where I could work during the day. In the evenings I hopped into the bus, pulled off the road somewhere nearby with a nice ocean view, and cooked my evening meal on a backpacking stove, then climbed into bed.

Shortly after returning to the States and purchasing my home on wheels, the Predatory Bird Program at UCSC announced that they were hiring people to monitor peregrine falcon nests throughout California. I jumped at the opportunity. Watching the dolphins had reinforced my commitment to becoming a zoologist. I felt most at home, most in tune with myself and the world around me, when I was watching animals in the natural world. I have always had a special place in my heart for peregrines, and watching them more intimately would broaden my background and experience. Besides, camping up in the mountains on my own would be an experience for which I would get paid, and I would have no expenses whatsoever. I brought the notebooks from Monkey Mia with me in my backpack along with a tent, cookstove, and a supply of peanut butter and crackers. So, under a mountainside peregrine nest, when the birds were inactive, I began the process of reading through our notes from Monkey Mia and writing reports and grant proposals.

When my work with the peregrines was finished, I went to British Columbia to watch orcas with a team of UCSC students who were conducting research. Orcas, though they look very different, are closely related to bottlenose dolphins. I figured it was important for me to see what other dolphins are like. The orcas were more impressive than any other creature I have laid eyes upon: giant dolphins dressed up in startling and elegant black-and-white costumes. The sound of their breathing carried far and wide across the mist-covered waters. Below the surface, their strident calls, which we monitored nearly continuously through speakers set up right in camp, echoed off the steep rock walls of the narrow strait.

Once during my two months with the orcas, I was lucky enough to get the chance to watch the whales visiting their "rubbing beach." For at least hundreds of years, whales have come to this particular stretch of beach to rub on the rocks, which are worn smooth as a result. As they approached, the whales sped up and began rolling and rubbing against each other, obviously looking forward to the rub with great anticipation. They swooped in among the rocks, and sitting on the cliff just above them, we could see white bellies flashing as they turned and wriggled and rolled and rubbed their backs and sides along the rocks and each other, like a bunch of puppies in a field of grass.

Orca watching was an education, but it didn't pay. I was determined to make enough money to get back to Monkey Mia, so I took up my parents on their offer and headed back east to live at home for a while, where I landed a job in a fine French restaurant. Wearing a formal black-and-white uniform that reminded me of the orcas, I learned to uncork exquisite old bottles of wine and to enjoy the delectable cuisine whenever possible. In a few short months I had gained a few pounds and also made just enough money to make it back to Australia.

In July of 1984, Sally Beavers had joined me on our return trip to Australia. A small, fair-haired woman, Sally was a student at UCSC with me, but it was while working together on the orca project that she and I had struck up a friendship. Besides being smart and fun, she was experienced in researching both orca and pilot whale behavior. Her experience, and a certain meticulous nature that I lacked but knew would be an essential asset, made her an ideal companion for a return visit to Monkey Mia.

We borrowed tape recorders and hydrophones with which to record dolphin vocalizations, and we were optimistically planning to beg, borrow, or steal a small boat that would allow us to follow the dolphins away from the camp at Monkey Mia. Though our heads were full of ideas, after we'd purchased our plane tickets, our pockets were close to empty. I had written a number of grant proposals, but nothing significant had come through as of our departure.

In order to save money, we arranged with a local Perth trucking company to hitch rides up the coast. Unfortunately, each truck could carry only one passenger, and the truck Sally was to ride in wasn't leaving Perth until two days after mine. She was hesitant to make her way out the final stretch to Monkey Mia on her own, so I would have to wait for her at the Overlander for close to three days.

It was excruciating to be so close to the dolphins and yet unable to proceed further. I pitched my tent in the bush back behind the station and passed the time reading, going for walks, and watching a sulfur-crested cockatoo that took an interest in my camp. Kangaroos thumped around during the night, and during the day the cawing of crows drifted on the wind from all directions.

I had ample opportunity to sample the Overlander diner cuisine. Like all stations in Western Australia, if not the whole of Australia, the diner offers up such delicacies as "Chiko rolls," "sausage rolls" (a processed meatlike substance wrapped in a fatty pastry), pies (more of the same served with "sauce," AKA ketchup), "pasties" (you guessed it, with a few chunks of potato and canned peas added), and the ever-popular "chips" (French fries, dripping grease). All day and all night for three nights running, trucks roared up the highway, laboriously shifting down, gear by gear, as they approached, grinding onto the gravel turn-off with hissing brakes, and then idling at a low growl while their drivers visited the facilities. Every one of them woke me up.

Finally, late on the third evening, Sally staggered out of one of these trucks, exhausted from the journey in a bouncing, smoke-filled cab. The next morning we hitched up with some travelers making a final pit stop at the Overlander before heading out to Shark Bay. They took us and all our gear right to Monkey Mia.

This was not the first or last time I had made my way from Perth to Shark Bay by means of hitching a ride with a truckie. Without a car and barely enough money to call our budget "shoestring," there were few options in those days. Later, with better financing and the more conservative lifestyle habits that seem to develop with age, we took to splurging on bus rides, bought a car, stayed in hotels if necessary, and paid to have equipment shipped up and down from Perth. I have never quite grown out of thinking of those things as luxuries, although now, in my middle age, I can hardly imagine doing what was both necessary and part of the adventure back then.

We finally arrived at Monkey Mia and were brusquely shown to our campground by Wilf and Hazel. As we walked through the campground en route to our site at the far end, we passed the familiar camps of many of the oldies we had shared the camp with two years earlier, including Bondi.

It had been nearly two years since Richard and I had camped in exactly the same place. The acacia bush had been chopped back, apparently by some camper searching for firewood, a rare commodity in this treeless environment. But the unimpeded view of the dolphins' bay was the same.

Sally and I headed off down the beach to see the dolphins. Holeyfin had a new baby, Holly. Joy, who was four by now, was still around and making occasional visits, but she didn't seem inclined to spend much time with people now that she was independent of her mother.

Holly was a gorgeous, energetic little female and, though she was only a few months old, had already taken to interacting with people far more than Joy had ever done. Much to everyone's pleasure, she would approach and, as she came within arm's reach, suddenly veer off and tear around in excited circles and then try again. She would place her tiny, perfect rostrum into your hands and rest it there, looking up as if waiting for a reaction. Holeyfin didn't seem to mind her antics and usually permitted people to touch her baby. But once in a while she would interject her own body between Holly and some person or chase Holly a little way down the beach, as if to reprimand her for being too forward with the humans.

Nicky and Puck still spent much of their time together roughhousing with the local gang of adolescent "guys": Snubnose and Bibi, Wave and Shave, Lucky, Pointer, and Lodent just offshore of camp. Now and then one of the Monkey Mia dolphins would break away from the fun and games and rush into the shallows, begging for fish handouts, excited and breathless from their energetic play, while the others hung just offshore, waiting.

The dolphins seemed to be spending more and more time in the Monkey Mia shallows, and we were busied with watching them much of the time. Still, we found ourselves embroiled in debates with Wilf and Hazel, various members of the oldies community, the local government officials, and others over just how best to protect the dolphins. Rumors circulated that Aussie tax dollars were being wasted to pay our way to Monkey Mia, where we lay about in the sun, drinking beer. Some of the misunderstandings that developed undoubtedly resulted simply from the generation gap between ourselves and most everyone else in the community. Others derived from the fact that we were foreigners. But mostly people smelled the winds of change, and it did not smell good to many. The dolphins and their admirers were the source.

One afternoon, just a couple of weeks after we arrived, Wilf Mason pulled his car up by our camp and let off a young bearded man with a backpack. "Here's where the Yanks go, you'll feel right at home." Andrew Richards set up his tent nearby and then came by for a visit. Back in his home state of Michigan, Andrew had heard a radio interview with Elizabeth Gawain and decided to come and see the dolphins for himself.

Over the next few weeks Andrew made himself indispensable to our efforts. He and I became fast friends and have remained so ever since. A shy, soft-spoken guy, prone to bouts of indecision, he is one of the most intelligent people I have ever known, possessed of both an encyclopedic knowledge of just about everything and the wisdom to make sensible use of his knowledge. He and I shared most of the experiences that are conveyed in this book.

With three of us working in shifts, Sally, Andrew, and I were able to do a thorough job of keeping track of the dolphins' comings and goings and still find time to work on setting up camp and seeking out a boat we could use. The two days Richard and I had spent out on the water in 1982 had convinced us that watching the habituated dolphins at Monkey Mia was but a small part of what we could achieve. The friendly, approachable, and numerous dolphins we had encountered offshore provided even richer potential for learning about the lives of wild dolphins.

The local fisheries inspector in Denham offered to let us use an old dinghy of his, but the moment we left shore, it began to fill up with water. One of the old-timers, an avid handyman and tinkerer, kindly offered to help us repair the boat. He spent hours over the course of a week trying to weld patches into the hull, but each time we took her out for a test run, she would leak from a different spot. Our excitement was turning to frustration. I began to think we were at the butt end of a bad practical joke disguised as a favor.

But the "bad joke" quickly went into retirement in early September when I received a letter from Bernd Wursing, relaying a message from the National Geographic Society. They had viewed our grant proposal favorably and given us $10,000 to support our work. I was incredulous at first. Somehow, though I had written the grant proposal, I had never really believed that they would actually give us real money.

Bernd had offered to serve as "principal investigator" (the person with a Ph.D.) on my proposal to the National Geographic Society, which had been a gamble for him. I was unproven, and he'd had to sacrifice his capacity to seek funding for other projects he was involved with. So it was no surprise that his letter closed with the comment "Now the onus is upon you to do a good job." I felt a knot of anxiety in the pit of my stomach. Dolphin watching was no longer just a fun pastime. This vote of confidence brought with it a certain weight of responsibility.

My first responsibility was to head back down to Perth for a shopping spree. A brand-new boat and motor, film, a big tent—one we could actually stand up in—and some simple furnishings. The trip to Perth took most of a week, and it was another week before everything arrived on the Shark Bay delivery truck. We had been at Monkey Mia for over a month already and we really hadn't even begun to do research. Still, we would finally be able to get offshore, get to know the rest of the dolphin population in the area, and expand our horizons beyond the Monkey Mia shoreline.

In our new little boat we ventured farther and farther from camp with each successive day. Our first objective was to learn to identify as many dolphins as possible. Each time we encountered dolphins, we took photographs of their dorsal fins, and then back at camp we pored over the slides, studying the distinctive shapes, scars, and nicks along their trailing edges. We gave each new dolphin a name. We could have given them numbers. Perhaps it would have been more objective, systematic, and scientific, but I knew I would never remember numbers. Besides, coming up with names was great dinnertime entertainment. We tried to give the dolphins names that reflected something about their natural markings. Bottomhook had a hook-shaped protuberance near the bottom of his fin. Trips had three little nicks in the middle of his fin. Chop was missing the top of his fin. Some were given names with more obscure associations. ED, whose fin was unscathed and hard to identify, stood for "enigmatic dolphin." YAN stood for "yet another notch," as she and three other dolphins with similar notches in their fins had all been confused for some time. We soon ran out of descriptive names, as there were so many dolphins, often with quite similar fins. So naming took on a more eccentric twist. A group of six dolphins that seemed to be together a lot were dubbed BJ, Bumpus, Bam-Bam, Biff, Bo, and Biddle. Mothers with babies were given names that went together: Yogi and Booboo, Zig and Zag, Tricky and Little, Blip and Flip.

Out on the boat, with dolphins surfacing and diving and moving around, identification was challenging. Fin identification is one of those skills, like mathematics, that some people seem to have a knack for and others just don't. Those lacking the knack could remain mystified by a fin they had seen a million times before. Visitors who accompanied us out on the boat sometimes thought we were pretending when we came upon a group of dolphins and reeled off a list of names.

Even for those who were good at it, identifying dolphins by their dorsal fins was definitely an acquired skill. In the early days, a typical notebook entry might read like this: "We just came across a group of about five dolphins. The one we were calling Notch but decided wasn't really Notch yesterday is here with a clean fin youngster. Also the one with the swept-back pointy fin with white scars near the tip that we have been seeing lately, and one that has two nicks evenly spaced at about one-third and two-thirds of the way down. That one also has a nick in the tail flukes and a scar on its left side, about halfway back from the dorsal fin. It's a round whitish mark about one inch diameter. Then there is another dolphin that could be Puck or Shave. That type of fin, but I don't think it's one of them."

After such a confusing encounter, we would have to wait for the slides to be developed before we could sort it out. Poring over hundreds of slides—a good windy-day activity—we developed a catalog of about two hundred different dolphins that we could identify. We learned to identify dolphins by their dorsal fins, much like recognizing a person by his or her face. It was immensely satisfying to encounter a group of dolphins and be able to identify each and every one. With that came the capacity to remember when and where we had seen each dolphin previously, what it was doing, and with whom. In short, we could build

To Touch a Wild Dolphin

a history of observations for each individual, something that had rarely been achieved with wild dolphins previously and yet was essential to understanding their lives. We were finally really making progress and getting to know the dolphins who were to become the core of our studies.

SHARK

BAY

In addition to getting to know the dolphins, we were getting to know the bay. The sights and sounds and smells of Australia's arid spinifex bush and the brilliancy of the bay's waters were becoming familiar. I felt less and less like an alien visiting another planet as the contours and colors of the landscape were becoming etched in my mind's eye.

Shark Bay proper is a huge area, about twenty-two thousand square kilometers, about half of which is water. Even the crudest map of Australia shows a glitch where the bay lies, just below the Tropic of

Capricorn. The bay is bounded to the east by the mainland of Australia, and to the west by the Edel Land Peninsula and a chain of barrier islands, Bernier, Dirk Hartog, and Dorre. Sticking out into the middle of this huge expanse is the Peron Peninsula. Monkey Mia is on the inshore side of the peninsula, and the village of Denham lies on the offshore side.

Occasionally I took time off from dolphin watching to explore the bay, packing a few camping essentials into the boat and venturing off into new regions for a couple of days. One such trip stands out in my memory as the time I first really caught a glimpse of the character of Shark Bay and realized it was not only the dolphins, but the place itself that drew me in.

With me on this trip were Debbie Glasgow and Nicki Fryer, two women I had met in the Monkey Mia shallows, among the dolphins. Both had developed a particularly strong attachment to the dolphins and had become astute observers of dolphin behavior. Nicki was a serious student of the spirit who spent her time going back and forth between an ashram in India and Monkey Mia. She was a slight, gentle, brown-haired woman with a shy but frequent smile. Though I sometimes found it hard to accept her belief that everything was part of a plan orchestrated by her spiritual leader, I appreciated the sincerity with which she sought a deep and satisfying meaning in life. We always referred to her as "Nicki-the-human" to distinguish her from Nicky the dolphin.

Debbie was a dancer and artist, with long braids that hung down her back or trounced about in the wind. She was possessed of a certain sparkle and energy that characterized her fun-loving, creative personality. Both Nicki-the-human and Debbie became long-term residents at Monkey Mia, doing odd jobs and living frugally in order to remain nearby the dolphins, and they often helped with our research.

Leaving Monkey Mia behind, I pointed our little boat up the coast of Peron Peninsula, along the edge of the flats, past Cape Rose, into Whale Bight, and then into less and less familiar territory up toward Guichenault Point. Even beyond Guichenault lies the top of Peron Point, visible from Monkey Mia only occasionally as a thin, wavering line on the distant horizon.

The weather was perfect—sunny and not too windy, with a few clouds building up along the horizon. I felt a little nervous way out there in no-woman's-land. It seemed safe enough when we were in close to the edge of the flats, not too far from shore. But when we crossed the open waters of Herald Bight, I felt exposed, vulnerable.

Debbie sat straight and quiet in the bow, braids streaming behind her in the wind. Nicki giggled nervously every now and then, apparently enjoying some personal reverie through the drone of the motor. I angled the boat into the waves and tried to keep track of the spray coming over the gunwales to make sure nothing got wet that shouldn't. We were all filled with the excitement and feeling of friendship that comes with anticipation of an adventure together.

Halfway between Herald Bight and Peron Point, we pulled into an attractive-looking sandy beach. Even though we would have to anchor the dinghy a quarter mile offshore owing to the shallow flats and wade to shore with our gear, the beach looked like an excellent place to camp for the night. The wind had dropped completely, and the sky was spotted with great gray clouds, some towering upward into the stratosphere—thunderheads that would likely bring rain before long. We acknowledged the possibility that we might be in for a very wet night. I felt a bit concerned about leaving the dinghy anchored offshore. It was our lifeline to civilization, and we would be in trouble if the anchor didn't hold and the boat was gone in the morning.

Just where we stepped ashore with our armloads of gear, a huge conch shell protruded from the sand, a luscious pinky orange color and completely unblemished. Usually by the time they make it to the beach they have been broken into pieces. This one must have been lucky, and it seemed like a good omen.

We set up camp, a fire pit, and tents, and then climbed up the ragged red cliff behind us to get a view. There were signs of goats everywhere. A bleached-out skull complete with horns lay in the red dirt, and streams of tracks flowed down the eroded cliff sides, crisscrossing each other en route to the water. Could these feral goats possibly be so thirsty that they are driven to drinking seawater? The only sounds were the singing of wedgebills and bellbirds and the subdued and rhythmic sound of water gently lapping shore. From the top we could see for miles out over Shark Bay on one side and back over the peninsula on the other. Like a mirror, the water reflected puffy clouds and streaks of glowing orange that shone through here and there. Several long, graceful arches of rain reached down, joining clouds to water and sweeping along across the horizon.

We decided to take a swim before settling in for the evening. With virtually no chance of running into another human soul, we stripped down naked and waded in. *Plip, plop, plip, plopity plopity plip plippity bshhhhhhhhhhhh.* It began to pour rain. Although the prospect of not being able to get dry was not too appealing, at least we would get a freshwater rinse, a rare treat around here. We danced around in the shallows, feeling as free and alive as I have ever felt.

After the squall passed, early evening sunshine peaked through the clouds, creating a double rainbow against the backdrop of steely gray cloud. Back at camp, we got a smoky fire going and crouched around it on haunches, giddy as we warmed ourselves and boiled water for tea, laughing and talking as the sun slipped in and out behind clouds and then finally dropped beyond the horizon. Out there we were delighted to be sticky from the saltwater and itchy with sand. We were beginning to blend into the landscape. Even the tiny camp of Monkey Mia seemed far away. The three of us were everything human in that vast wild place.

In the morning we set off for a hike, traversing sandy beaches, over rocks, up and down cliffs. When we got too hot we jumped into the water, then tied our sarongs around our heads to keep off the midday sun (which resulted in sunburned bottoms). We didn't walk together most of the time, but rather each settled into her own rhythm and time, preferring to tune in to the surroundings rather than conversation. Now and then we converged and sat together, sipping water from our bottles and comparing notes on what we had seen. Birds reeled overhead, and several times during the day we saw dolphins just offshore. I wondered if they were dolphins we knew, way out here, or strangers.

Around midday we came to a beach strewn with large rocks, shaggy with seaweed, and covered with tiny oysters. Using small rocks, we pounded the oysters open and gouged out the tiny sweet morsels with bare fingers. We began to fantasize about living out here. Maybe we should just stay and live off this land. No, not with the lack of freshwater. That would drive us back to civilization if nothing else did. Still, Shark Bay affords a level of wild isolation and beauty that is simply unsurpassed in my experience. It has a unique and vibrant character that at first is blinding. You think you have landed in a flat and featureless dull place. Before long, you learn to see it, and it quickly becomes an addiction. Its colors and smells and sounds take up dwelling in your heart. It is both gentle and harsh, generous and meager, bold and subtle.

Back at camp that evening, I scaled the cliff for a sunset view: the land, crouched low under a huge sky, sweeping out to the horizon, thinning and then disappearing into the Indian Ocean. As is often the case in Shark Bay, it was difficult to tell where land ended and sky and ocean met. The wind blows persistently, sculpting brilliant red dirt, cream-colored sand, and sage green scrub into shapes that offer it the least resistance. Sand rolls in drifts, beaches shift, clouds of red dust rise, twist, and resettle, bare craggy cliffs, stained white in places with the droppings of falcons and white-breasted sea eagles, jut out against the blue sky. Stubby acacias and wattles never grow more than about eight feet tall, and most are much shorter: tough, gnarly, charred, and twisty things built to withstand the harsh, dry climate.

The first time I traveled to Shark Bay, I felt as though I had truly arrived at the ends of the earth. In the town of Denham, a few buildings lined up along one side of the main road faced out on a vast watery expanse. The Edel Land Peninsula and Shark Bay's barrier islands lay in the distance, a thin broken strip along the horizon constituting the only land between me and Madagascar. The whole place was illuminated by a strangely brilliant and unfamiliar light.

Shark Bay is the westernmost point on the continent of Australia. It was here that the first European explorer set foot on the southern continent. For a long time, Australia had been hypothetical in the minds of European explorers. They suspected it must exist, that there must be something there in that big expanse of southern ocean. But with only relatively crude navigational tools, a series of previous attempts to find the continent had failed. A few hardy souls had sailed so close that they were virtually within earshot, yet, unaware of their proximity to that strange new world, after months of hard sailing from Europe, they had turned wearily back home.

A Dutchman, Dirk Hartog, was the first to land. In 1616 his boat, the *Eendracht*, pulled ashore on what is now known as Cape Inscription, a bare, windy promontory on the island that now bears his name, one of Shark Bay's barrier islands. He marked the spot with a small pewter plaque. But Dirk Hartog's plaque did not rest easy. In 1697 Captain de Vlamingh, another Dutchman, stopped at Cape Inscription, pulled down Dirk Hartog's plaque, and nailed up his own. Years later the original plaque was reinstated. Meanwhile, in 1722, François de St. Allouarn landed at Cape Inscription and made a formal claim of territory for the French, burying two French coins and a parchment in a bottle, as well as the remains of one of his crew members. It is hard to imagine how this barren spot inspired such competition.

The bay was named Sharks Bay by an Englishman, William Dampier, in 1699. Two French explorers, Captain Baudin aboard the *Géographe* and Captain Hamelin aboard the *Naturaliste*, first charted the bay and its waters in the early 1800s. In 1827 the first British settlements in Western Australia were founded farther south at King George Sound (now Albany) and Swan River (Perth), and eventually, the whole of Australia came under British rule.

Of course, Australia's history began not with Europeans, but rather, some forty thousand years earlier with the Australian aborigines. Shark Bay was home to the Nanda and Mulgana tribes. The oldest remains of aboriginal settlement in Shark Bay, middens (piles of marine shells, crab claws, mammal bones, and stone artifacts), date back about five thousand years, comparatively recent. But Shark Bay has not yet been subjected much to the scrutiny of historians and archaeologists; there may well be older secrets that still remain unrevealed.

Shark Bay's dry landscape preserves things well and is adorned with bits and pieces of its past. I used to wander aimlessly through the scrubby brush. If a patch of dense prickly acacia blocked one direction, I took another. If the massive web of a golden orb spider or two stretched across my path, I sought another way. I always waved a stick in front of me to clear the spiderwebs, which can be so massive and tough that it takes some strength to push through them (and they snap alarmingly when broken). Turning and twisting my way with only a vague sense of purpose, I often came across signs and remains from another time.

One of the most common finds were the old white clay pipes used by the Chinese and Malay pearlers. It was usually when I stopped and sat down in the middle of the bush that I would find things. Stilling my breath and my mind, listening to the sounds of the bush, and attuning my senses to this different perspective, I would catch sight of a flash of white alongside the gnarled trunk of an acacia, and sure enough, it would be a pipe. Here in the middle of what seemed the most remote and unlikely spot, someone else had sat long ago, relaxing with a bowl of tobacco. He had no doubt been listening to the same sounds of wedgebills and bellbirds, pushed his way through the same maze of spiderweb-laden spiny brush, and watched the same variety of ants milling around on the red dirt.

Another common find were bits of a dense, shiny, cream-colored stone. This always struck me as odd because I never saw any of this kind of rock occurring naturally in the area, and the bits I found lay surrounded by dissimilar rock and sand. They looked vaguely to my untrained eye as if they had been chipped and worked by human hands, and I found myself pocketing them and taking them back to my collection of treasures at camp. Later I learned that there is a quarry where the aborigines used to collect this Yarringa chert, specifically for making stone tools. Bits of the worked stone originating from this quarry are found scattered over much of the Shark Bay area, sometimes hundreds of miles from the quarry.

Looking around at the landscape, one could easily visualize the aborigines. They survived and even thrived here, adapting their knowledge and habits to the particular richness of this place; they caught fish, shellfish, dugong (a sea cow, similar to the manatee), turtles, lizards, and birds. The aboriginal eye probably found Shark Bay reasonably hospitable. To the European eye, however, the land appeared hard, unforgiving, barren. No rolling meadows filled with grass and flowers and grazing deer. No shady trees or bubbling brooks. The disappointment and despair of those early European explorers is reflected in such place names as "Hopeless Reach," "Disappointment Reach," and "Useless Inlet." In those days, scurvy and malnutrition spelled the deaths of many sailors, and most were desperate for fresh water and familiar plant and animal foods with which to provision their ships. But this harshness is in part a mind-set, born of habit.

The waters of Shark Bay are surprisingly shallow, and early explorers, lacking any charts, no doubt ran aground and had to backtrack through a maze of channels and sandflats in their attempts to enter the bay. The average depth is only about thirty-five feet, but about a quarter of the water area is just two to three feet deep. Sand- and seagrass-covered shallows extend offshore along much of the coast, in many places for a couple of hundred yards. It is all but impossible to bring a boat with any draft whatsoever close to the beach. Once when we visited Faure Island, a small island about nine miles off Monkey Mia, in a forty-foot single-hulled sailboat, we had to anchor a good three-quarters of a mile from shore, out on Green Turtle Flats. We waded in on a good low tide in water only knee-deep at best. But as we explored the island, the tide was coming in rapidly. We started back out to the boat as the sun was getting low and quickly found ourselves in chest-deep water and surrounded by sharks.

The shallows of Shark Bay can also be a problem for dolphins. Once during a follow of the dolphin Sicklefin, we found ourselves coming into water barely deep enough for our little boat. We persisted and before long found ourselves in a labyrinthine maze of sand and seagrass banks, many of which were simply too shallow to cross on the present tide. We kept expecting Sicklefin to turn around and head out into deeper water, but he continued along, his back and dorsal fin continuously exposed and humping up and down as he wriggled along. Finally he came to a flat that was barely covered with water. Instead of seeking an alternative route or turning back, he headed straight up onto the bank and bounced along, seal-like, on pectoral fins and belly, right across the bank and into deeper water on the other side.

The shallowness of Shark Bay is also part of what makes it such an ideal place to observe dolphins. In many places, even when the dolphins dive all the way to the bottom, they are still visible (depending on the water clarity). On a clear water day, as one drives slowly along, a wondrous underwater world rolls by. A turtle looks up from its resting position on the bottom, a ray darts off, kicking up a plume of sand as the boat approaches, a volute with its geometrically patterned shell makes tracks on the sand. Great patches of different varieties of seagrasses pass by, like a tropical forest in miniature. When dolphins are near the boat on such a day, every little scrape and scratch on their skin is visible, as is every subtle little change in position: a flick of the tail, a twist of the pectoral fin, a roll of the eye. They seem to float effortlessly, suspended in the water against a backdrop of sand, seagrass, and the shiny silver-gray bodies of other dolphins.

In the mid–nineteenth century, Shark Bay was the site of a thriving industry in bird guano, thanks to some large breeding colonies of cormorants and other marine birds. Shortly thereafter, the pearling industry followed, and Monkey Mia became the site of a major pearling effort. Chinese and Malay pearlers settled around the bay, and today the current inhabitants of Shark Bay are a mixture of aboriginal, European, Malay, and Chinese blood. One theory about how Monkey Mia got its name holds that the Europeans referred to the Chinese pearlers as "monkeys," while Mia means "home of" in the local aboriginal language. Huge piles of oysters could be dredged up in a single day. These had to be opened, one by one, in search of pearls. The oysters themselves were collected in huge vats and allowed to rot, then boiled down in yet a further effort to find any remaining pearls. The smell of boiling rotten oysters ("pogey") must have permeated the entire bay in those days.

On one of my bush "walkabouts," I found a small camp, perhaps left by some of the old Chinese or Malay pearlers many years ago. I had wandered way out into the bush and eventually found myself in a thicket of scrub so dense and impenetrable that I had to get down on my belly in the sand and squirm along underneath to get through. There in the midst of this thicket I came across a little clearing with a collection of rusted, old-fashioned metal pots and trays and the remains of a fire pit. A casual look around, and I found a couple of pieces of white clay pipe and then a small, very ornately designed button, lying on the sand. The design was distinctively Asian.

Fishing ultimately became the mainstay of the bay. Snapper and grouper caught on wetline, and mullet and whiting caught in nets, were good value, as were prawns and scallops dredged up from the bottom, a practice that destroyed large areas of seagrass beds and was later restricted.

In the early twentieth century a "lock hospital" was established in Shark Bay. Two facilities were built, one for men and one for women on Bernier and Dorre Islands. Aborigines from northwest Australia infected with venereal diseases or leprosy were shipped to these horrendous prisons, where they were locked up and left to waste away, isolated on the remote islands, far from their families and relatives, sick, and with no hope of escape.

Shark Bay is uniquely situated at the point that marks the southern boundary for organisms with a northern, desert climate distribution and the northern boundary for southern, more temperate climate species. It is the zone of overlap, the area where two quite different ecotypes meet and mingle. Driving up the west coast, one begins in the rolling green farm country just north of Perth. Flocks of red-and-black cockatoos flap noisily over tidy vineyards. Grassy green pastures are dotted with cattle, sheep, and the occasional kangaroo or emu. Gradually the landscape transforms. The trees shrink in stature as banksias, acacias, wattles, and spinifex replace the tall eucalyptus. Expanses of arid brush, sparsely covering pale pink and brilliant red sands, extend to the horizon on all sides, most of it untrammeled by any human activities. The sky grows larger and larger, demanding more and more consideration as the landscape narrows.

The beauty of Shark Bay is not the sort that necessarily knocks your socks off at first glimpse. One must look carefully to see it: in a kangaroo print molded into deep red sand alongside a cluster of deep purple flowers and sage-green foliage, colors both subtle and startling; in the soaring of an osprey riding the updraft along an oceanside cliff; in the most luminescent, spectacular, profound sunsets I have ever witnessed.

This magnificence has not gone unnoticed by the world at large. In 1991 it was designated a "World Heritage Site" by the United Nations. Parts of the bay have been granted marine park, national park, and reserve status. It is home to the world's most extensive seagrass banks, the Wooramel Banks. At least twelve species of seagrasses provide refuge for juvenile fish, prawns, a few species of sea snakes, four species of turtle, and the remarkable leafy sea dragon, among other creatures. Seagrasses also provide fodder for one of the world's largest remaining populations of dugongs.

The seagrass beds are so extensive that they actually form a baffle in some areas, limiting the circulation of seawater and resulting in areas where the water is as much as twice as salty as normal seawater. One such place is Hamelin Pool, a big shallow basin scooped out of the elbow between Peron Peninsula and the mainland. Hamelin Pool is so salty that few creatures can tolerate it. One organism that thrives, however, is the stromatolite. Stromatolites are colonies of cyanobacteria (formerly called blue-green algae.) As they grow, the tiny creatures trap and bind sediments that are deposited to form mounds. Only a thin layer on the top of the mound comprises living cyanobacteria, perched atop the deposits of their ancestors. There are several kinds of stromatolites made by different assemblages of cyanobacteria, each forming its own characteristically shaped mound: "coloform," "gelatinous," "smooth," "pincushion," "tufted," "pustular," "film," "reticulate," and "blister mats."

Stromatolite mounds grow incredibly slowly: about 0.3 mm per year. Some of the mounds in Hamelin Pool are so large, they are estimated to be thousands of years old. These ancient stromatolites are the descendants of an even more ancient legacy. In the earth's fossil record, stromatolites are among the earliest life forms to appear. Paleontologists had long been confused by fossil stromatolites, dating back about 3.8 million years to the Precambrian, which appeared at first to be some sort of rock. They were recognized as fossilized life forms only after modern, living stromatolites were discovered. Now, stromatolites are limited to only a few places on the earth's surface, and Hamelin Pool is one of their last strongholds. Biologists think this might be precisely because of the saltiness of the water. Animals that would otherwise graze on the stromatolites simply cannot survive in Hamelin Pool. With no predators, the stromatolites hang on. Now their gravest threat is curious tourists. A few careless footsteps can destroy what has taken thousands of years, in some cases, to build.

Another Shark Bay oddity is the coquina shell: a tiny, beautifully symmetrical bivalve, *Fragum erugatum*. The coquina also benefits from its ability to thrive in the hypersaline waters where predators and competitors dare not. And thrive they do. Shell Beach, along the edge of Lharidon Bight, consists of mile after mile of the tiny shells, washed ashore and bleached pure white by the sun. These shells are piled fifteen feet deep along parts of the beach. Where they pile up, they tend to compress into a hard cement. People in the Shark Bay region chop blocks of the shell cement to use in building construction: a good idea, given the lack of timber. The coquina block buildings, like the waterfront "Old Pearler" restaurant and the church in Denham, not only look beautiful, but are also well insulated against both heat and cold. Even the streets of Denham used to be paved in coquina shell.

Besides the stromatolites, the coquina shells, the seagrass beds, and dugongs, turtles, rays, sharks, and, of course, the dolphins, there is the sandhill frog, the only frog species that never, at any stage in its development, inhabits water. In fact, it spends its days burrowed into the sand—hot desert sand, that is—where it broods its young, which hatch directly from egg to frog, skipping the usual tadpole stage.

The unusual shape of Peron Peninsula, which constricts to a very narrow bottleneck near the bottom, creating what amounts to an island of land barely connected to the mainland, makes it an ideal site for the ambitious "Project Eden." The goal of the project is to fence off the peninsula at the bottleneck and then get rid of the introduced pests from the top of the peninsula. With a combination of natural regeneration and careful reintroduction, the hope is to re-create a natural Australian ecosystem, teeming with native marsupials like the western barred bandicoot, the red-tailed phascogale, and the rufous wallaby, all of which used to inhabit the peninsula.

The Shark Bay bush is full of birds. Groups of brilliant little fairy wrens in shades of iridescent blues flash by. These birds usually form groups with several females and a single male. I once saw a nesting pair feeding a baby cuckoo. Cuckoos are nest parasites. The parents lay their eggs in the nests of other species (the wren, in this case), who then are fooled into raising the young cuckoo instead of their own young. How they can be so easily duped seems mysterious to me. The baby cuckoo is several times larger than any baby fairy wren would be. In any case, when a friend and I approached this parasitized nest, the little male wren hopped down onto the ground and distorted his little body, hunching over and pushing feathers out, and then he ran along the ground in an odd manner. This was the "mouse display." Like the "broken wing" display of some other bird species, the "mouse display" is thought to distract predators like snakes who always have an eye out for mice.

Noisy, scolding, chattering crowds of white-browed babblers pass by, poking and prodding at everything in their path with their long, downturned bills. These birds, like the wrens, are quite social living in large groups, probably fathers and brothers and some females. Once I spotted a flash of red as a babbler flew by me. I was not sure, but it seemed possible that the bird was sporting a leg band. When I reported this to a friend who was researching birds in the area, she said her father, also a bird researcher, had marked quite a few babblers many years earlier. She went to the spot where I told her I had seen the marked bird and, sure enough, confirmed it. The bird had been marked ten years earlier by her. This was probably the longevity record for a wild babbler.

Finally, a common denizen of the scrub in these parts is the rather drab-looking, chiming wedgebill. But what he lacks in looks, this little bird makes up for in song: a five- or six-note phrase sounded in hauntingly beautiful, bell-like tones. The bird seems to need a warm-up period in order to build to the full effect. As he starts out, his mouth moves, but the sound is faint, and like a ventriloquist's voice, it seems to be coming from a distance—somewhere, you can't tell, even when standing right next to him. But after a few runs through the phrase, the song builds to take on a ringing, piercing, reverberating quality that seems highly unlikely coming from a little gray, tufted bird. The four-note stanza ("did you get drunk") changes now and then, as emphasis is moved from one note to another in different renditions.

Denham, the main settlement in Shark Bay, is a town of contrasts. Just next door to the beautiful little coquina-shell church, with its distinctive Shark Bay charm, is a building typical of the new style: a square box with aluminum siding and a corrugated tin roof, all dull and stained by the onslaught of red dust, plopped onto a little fenced-in square of bare dirt.

Some people here seem to have coped with their isolation by extending the warmest hospitality to outsiders, while a few are suspicious and distant. In the past ten years this little village has undergone an incredible transition from an isolated little settlement, clinging to a hard and often meager way of life, to one of the major tourist destinations in Australia.

People often claim that pet owners look like their pets—the fat, scowling woman with the bulldog on the leash, the dandied-up dude with his impeccably behaved and highly tense afghan, the ditzy little woman with gaudy bangles, a frizzed-up hairdo, and a poodle. I think the same extends to landscapes. People take on a certain character, perhaps a result of breathing its air and drinking its water for many years. Shark Bay people are tough and leathery from the sun, salt, and wind. Their skin is reddened by the dirt that works its way irreversibly into the pores. Like the land, they can sometimes appear harsh or unsophisticated. But look a little longer and they are a fascinating study.

Far more diverse in looks, character, experience, skill, and manner than the homogeneous townsfolk of my experiences elsewhere, Shark Bay people are not accustomed to luxury. They have not grown up with access to the latest in technology and comfort. Instead, in their isolation they have had to rely on their own skills and wits, handling boats, dealing with extremes of weather, driving rough tracks up to the axles in red dirt, catching and processing tons of fish, taking apart motors and putting them back together again, constructing their own buildings. Lately they have added to their repertoire the tending of tourists: cityfolk who arrive in this isolated place with punctured tires, failing engines, and ignorance of the local ways and waters. Although the development and the changes that have occurred as a result have not come easily, they are but another moment in the long, rich, and ongoing history of the region.

Debbie and Nicki-the-human and I spent three days at our camp up at Herald Bight. By the end of our stay we were as brown and wild as our middle-class European backgrounds could possibly permit. Reluctantly we loaded everything back onto the dinghy, having timed our departure around a high tide to ensure that we could get out of the shallows. We turned back toward Monkey Mia and drove in silence, letting our return to civilization seep in slowly and gently so as not to be too much of a shock. Crossing open water no longer made me uneasy. We had found our own wits and capabilities quite sufficient out here, confidences were restored, and I felt as though I were humming in tune with the harmonies of the bay. All was well. We rounded Cape Rose and were met by a group of dolphins, Lucky, Pointer, Lodent, and a couple of young males we didn't know. They greeted us by riding at the bow and then broke away to resume their activities.

I realized just how deeply this place had affected me. The company of dolphins, the brilliancy of light and color, the constant company of wind and sand, the sparkle of sunlight reflected off water, the now familiar songs of the birds, all had permeated my being and taken hold. Farther into Red Cliff Bay we were met by Nicky, Puck, and Crookedfin. They rode our bow right into camp. Welcome home!

FOLLOW

A WILD

DOLPHIN

As my sense of belonging and appreciation for the place grew deeper, so did my determination to really get to know the Shark Bay dolphins and to contribute in some significant way to humankind's understanding of these creatures, perhaps even to inspire people to protect and revere them. And my way, the way I have always felt to be most useful, reliable, and honest, was through science.

People make their mark on the world in many different ways, all of which I believe contribute to an interesting culture. Some write poetry, paint pictures, express their emotions through art and music and dance. Others make their mark on the world by doing good deeds. For me, there is something especially exciting about establishing facts, elusive though they may be, contributing to the truly awesome progress of scientific discovery. It strikes me as incredible that not too long ago we didn't know that the earth was round or that it was one of many planets in a galaxy within a universe full of other galaxies, or what causes weather, or the molecular constitution of water or air. We didn't even know of the existence of bacteria or how to cure most common diseases. Only very recently have we discovered genes and begun to understand some of the molecular basis of evolution, heredity, and behavior.

Generation after generation, humans have honed and refined and developed means of discovery in a massive and successful effort to comprehend. Though the progress is slow, and the argument can be made that it is impossible to be truly objective, I find it impossible to deny the value of the scientific method. Without science, we would have no reasonably objective basis whatsoever for understanding the world. We would have nothing but hearsay and personal bias. As Ken Norris used to say, "Science is a system of rules to keep us from lying to one other."

It was with this general mind-set that Richard, Andrew, and the other researchers who later joined us began to develop what we considered to be reliable methods for studying the Shark Bay dolphins. Once we could identify individuals, we would really be able to study their ways. Some of the basic questions we wanted to answer—What sorts of groups do the dolphins form? Where and how far do they travel? What do they eat, and how do they catch their prey?—seemed relatively straightforward. Other questions—What sorts of relationships do they have with one another? How do they use their brains in their natural environment? Are their sex differences in behavior? How do youngsters develop and mature into adults?—would be far more difficult and time-consuming to address.

This was going to be several lifetimes of work. Answers would come, not immediately, but in their own time. We needed to think through the questions and what information we would need to know in order to provide answers. Then we could develop ways to observe and collect the kinds of systematic data that would allow us to answer these questions to our own satisfaction and to the satisfaction of the scientific community.

Weather permitting, we usually filled our days with dolphin watching, beginning with the hectic process of preparing for a day on the boat. Did we remember the spare batteries for the tape recorder this time? Do we have enough tapes for the day? How much gasoline is in the tank? Did we label the film canisters? Is there water to drink in the bottles? Sunscreen, sunglasses, hats, windbreaker? The checklist got longer every day, and still we sometimes found ourselves way out on the bay with an interesting group of dolphins only to discover that we had run out of film or fuel. After we'd hauled everything down the beach and packed the boat, it always felt exhilarating finally to jump aboard and head out into the wild, sparkling bay, full of anticipation about what the day would bring.

On one such day, Andrew and I set off together.* He has smeared white zinc oxide all over his nose, lips, and cheeks, and I tease him about the way he resembles a clown. My darker skin doesn't burn as easily, so I allow vanity to prevail for the moment.

We head out to the northwest, picking up speed and enjoying the coolness of the air blowing past, scanning the water surface for dorsal fins. Sometimes the bay seems to be teeming with dolphins, and every time we turn around there is another group to check out. On other days we drive and drive, searching endlessly for just one dolphin to watch.

---

*The events described here are presented as having occurred during a single day's work, though in fact they are compiled from several different days.

When the weather is calm and the bay turns glassy, as it is today, the best strategy is to stop the boat, turn off the motor, and listen for the sound of dolphins breathing. Sound travels far across the water surface on these days, so we can hear and therefore locate dolphins over an area of a couple of square miles. I'm always surprised at how many dolphins there are. It is not uncommon to find ourselves surrounded, north, south, east, and west, by small groups and solitary individuals, scattered around, going about their business.

When conditions are less favorable, the bay can seem devoid of dolphins altogether. The dolphins, though, probably don't have this problem. Underwater they can be quite loud and, with their keen sense of hearing, probably listen to each other and keep in touch, even over large distances. I wonder how our conception of their world would change if we could see with their eyes and hear with their ears.

Andrew and I stop to listen and hear the characteristic *phoohoop*— exhale and inhale in one gesture, as a dolphin breathes. It sounds close, but we see nothing. Finally, way out, along the smeared and wavering horizon close to a mile north of us, I can just make out a dolphin's tail swinging up into the air and then submerging. Bringing the tail flukes up into the air is an indication that the dolphin is heading straight down toward the bottom. Tail-out-dive foraging, one of several foraging strategies we came to recognize, is probably the most common dolphin activity in Shark Bay. Usually on their own (or with their babies, for mothers), a tail-out-dive foraging dolphin travels slowly along the surface, takes a few breaths, and then swings his or her tail up and goes down, remaining below for about two minutes before resurfacing.

Presumably the dolphins are hunting for fish that are located at or near the bottom. The dolphins frequently change direction underwater, so it is one of the hardest activities to keep track of. A lone dolphin is always harder than a group, and a dolphin traveling erratically is even harder. We feel compelled to try to determine who it is, maybe even get a photograph of its dorsal fin if necessary. Every time we encounter a dolphin (or a group of dolphins) we fill out a sighting form, noting down the identity of the dolphin(s), location in the bay, activity, and other information that might be useful. This can take a lot of patience and effort, particularly with a hard-to-approach dolphin who is tail-out-dive foraging.

From south of us several hundred yards, Andrew and I hear the multiple staccato breaths of a tight-knit group of three or four dolphins as they pop up to the surface together. After that first breath they seem to breathe less quickly and less synchronously. They are moving slowly and, after a couple of minutes, disappear below the surface again. The pattern of their breathing and travel tells us that they are probably resting.

Resting dolphins form tightly knit groups that spend several minutes below the surface between breaths, cruising close to the bottom, then rising in unison to the surface to breathe. Several breaths later they will all disappear again as they dive. Dolphins also rest by snagging: hanging at the surface with their foreheads exposed and the rest of their bodies suspended just below the surface. When groups of dolphins snag together they line up rank abreast and just hang in a row, foreheads exposed and glinting in the sun. They look like a row of sausages, "snags" in Aussie slang, and hence the name.

Like a student struggling to be attentive through a boring lecture, a snagging dolphin often has at least one eye partially shut and seems just barely alert enough to remember to breathe. Most mammals, like ourselves, don't have to remember to breathe; breathing is under involuntary control; and even when we are asleep or under anesthesia we continue to breathe. Not so for dolphins. They must be conscious of each breath. This makes sense for an animal that must rise to the water surface and get its nose out into the air for each breath. Anyone who has gone snorkeling knows what this is like. You can't breathe whenever the urge strikes, you have to time it just right.

Voluntary control over breathing in dolphins was discovered initially by researchers working on brain anatomy. They tried to anesthetize their dolphin subjects in order to apply electrodes to their brains, but as soon as the dolphins succumbed to the anesthesia, they stopped breathing and died.

Perhaps because of the need to maintain enough consciousness to remember to breathe, dolphins do not sleep in the same way as most other mammals. They do not experience REM sleep and the brain wave patterns typical of that state. Dolphin brain wave activity during sleep alternates between the two different hemispheres of the cerebrum, apparently leaving one side of the brain alert enough to juggle the logistics of breathing.

The only obvious movement in a snagging dolphin is the occasional tiny corrective motion of a pectoral fin. Maybe they are just staying within easy access to air, sunning themselves, or listening. Occasionally a snagging dolphin will stretch, first arching its head up, flexing its back so that face and tail point upward, and then reversing the motion, tucking its head down and below the water surface, arching its back so tail and snout point down. Males sometimes display an erect penis when they stretch. Perhaps they're dreaming.

Watching dolphins snag is usually pretty uneventful. Their sleepy demeanor, the slosh of water against the side of the dinghy, the gentle rocking, and the heat of the sun all conspire to make us feel sleepy, too. When they stretch, I am made suddenly aware of my cramped position on the hard metal bench of the dinghy.

Wild dolphins I have watched outside of Shark Bay never seem to stay still. By comparison, the Shark Bay dolphins seem to spend a lot of time snagging. I wonder if it may have something to do with the very high salinity of the water in Shark Bay, which, because of its greater density, makes it much easier to float. After years of swimming in salty Shark Bay, I moved to Michigan and jumped into a freshwater lake to take a swim. I sank like a lead weight and had to work hard to stay afloat. Maybe the dolphins in Shark Bay snag just because that is where they end up when they completely relax, being just slightly positively buoyant in the dense salty water.

Andrew and I decide to follow the resting group, which, after a few breaths and a bout of snagging, has gone down below the surface again and is not visible. As we putter toward the place where they went down, I concentrate on trying to make sure I see where they come up again. It is remarkably tricky to estimate distances on the water. Many times we have driven up to where we thought a group of dolphins was during their last interval at the surface, only to have them come up way behind or ahead of us.

Hoping to avoid disturbing them with our approach, we slow way down as we get near the spot where the dolphins went down. With patience and perseverance, we won the acceptance of most of the dolphins we encountered regularly by approaching them slowly and driving carefully around them. We spent many hours with most of them, and they must have learned to recognize the sound of our boat. Occasionally we encountered dolphins who seemed shy and frightened of us. They would turn away repeatedly, dive evasively, or even take off at top speed when we approached. But if they were in the company of dolphins we already knew, and who were comfortable around us, they seemed to take their ease as well. Seeing their companions approach our boat in a relaxed manner encouraged them to trust us. The dolphins' acceptance of our presence ultimately was perhaps the most important factor that allowed us to make the observations and discoveries we made.

Unfortunately, not all boaters are so careful. The worst offenders inevitably are driving enormous powerboats painted with bright racing stripes and emblazoned with names like *Raider* or *Pursuit*. Oblivious of all but the thrill of their ride, they plow directly over the dolphins, whom they apparently don't even see, leaving the dolphins and us disgruntled and bouncing in their wake. The dolphins dive out of their path and under the boat. Underwater, the noise created by these huge motors is deafening, shattering the peace.

With the many shallows that are characteristic of Shark Bay, these joyriders, being unfamiliar with the topography of the bay, are likely as not to run aground on a bank sometime during their adventure, damaging their boats as well as the delicate ecology of the seagrass banks. Many of the shallow banks directly in front of Monkey Mia are crisscrossed with scars from boats that have cut across at low tide, chewing up a swath of seagrass with their propellers. The remaining bare sandy paths can take years to regenerate their seagrass covering.

Today the bay is peaceful, with just a few small boats puttering around, mostly fishermen. The dolphin group resurfaces not far from the footprint, a patch of flattened water marking the spot where they submerged, left by their previous surfacing. When we get within about fifteen feet of them, the whole boat suddenly begins to wobble a little from the force created by a bowrider's tail flukes sweeping through the water just in front of us. It's Peglet, Square's two-year-old daughter. I lean over the bow to whistle my usual greeting to her. She is lively and excited, wiggling and zigzagging back and forth as if preparing for a high-speed joyride of her own, comical given that we are barely moving at one knot.

As Peglet rides the bow, she tilts her head slightly to look at me first with her left eye and then with her right. Leaning over the bow, I am just inches from her as she glides along just under the water surface. When she comes up for a breath, the salty, slightly fishy-smelling spray of her breath flies back in my face. Such a good view of her body, up close and in clear water, provides an excellent opportunity to assess her physical condition. She looks good, plump and rounded, not sunken in behind the head or along the peduncle (tail stock), as happens to sick dolphins.

Like so many dolphins in Shark Bay, Peglet bears the scars of a close encounter with a shark. She is still young, so her encounter must have occurred when she was an infant, given that it is already well healed. It's hard to imagine what it must be like to swim around in the murky waters, particularly at night, with so many sharks. It's even harder to imagine what it is like to get bitten. Some dolphins bear huge scars where a large portion of their body was engulfed in the shark's jaws and chunks of flesh ripped off. I am amazed that they survived these injuries. Many must not.

Peglet keeps glancing up at me with an inviting look. She whistles, emitting a tiny stream of bubbles with each whistle. Is she greeting us? I whistle back. She rolls upside down and dives away from the boat, but not before I catch a glimpse of her belly. She is only a youngster, still dependent on her mother, so her belly is still pure snowy white, without a single speckle. The tiny mammary slits lying alongside of her genital slit reconfirm our assessment that she is female. We had discovered, much to our delight, that we could often determine whether a dolphin was male or female. In females the gap between the genital and anal slits is small, and the genital slit is flanked on either side by two small mammary slits, within which lie her nipples. Males have a long genital slit (sometimes with a bit of a bulge to it), a larger gap, about an inch or two, followed by the anal slit. With the more friendly dolphins, we often had an opportunity to view their bellies when they rolled upside down while riding at the bow of our boat. We made every attempt to encourage them to ride, fun for us and for them. Over time, even most of those who would not ride at the bow could be sexed either because they had a baby or because we observed them with an erect penis. The ability to determine sexes so easily was unprecedented in studies of wild dolphins, and it proved essential to making sense of most everything we later discovered about dolphin society.

Peglet heads back to the rest of the group, predictably consisting of her mother, Square; two other adult females, Fatfin and Tweedledee; and her older sister, Squarelet, now an adolescent. These females are among the most consistent female friends in the bay. We don't know the exact relationship among Fatfin, Tweedledee, and Square and her daughters, but it's possible that Fatfin and Tweedledee are Square's sisters and therefore Peglet's aunts.

Andrew and I decide to do a focal follow on Square. We will try to stay with her for as long as possible. On a calm, clear day, that may be many hours. In the early days of our dolphin watching, we had made a point of trying to locate and identify as many dolphins as possible. When we encountered a group, we would determine who was present and assess what the general group activity was. If the group was interesting to us, we would stay with them for a while, but eventually we would set off in search of more dolphins. The data we collected in this manner were useful for answering certain types of questions. For example, because we encountered so many different groups of dolphins, and assessed who was present at each encounter, we could say something about the size and composition of groups and about which dolphins tended to be together. But one big problem with our approach was that we watched dolphins only when they were doing something we deemed interesting. Because of this, we had a biased impression that dolphins were always socializing, since that was the activity that most interested us. We learned to see some of the more dramatic and obvious facets of dolphin behavior but had little sense of the normal ebb and flow of activities: their biological rhythms, how much time they spent with one another or engaging in different activities.

Those insights came when we adopted the practice of doing "focal dolphin follows." All this means is that at the beginning of the day we selected one particular dolphin as our "focal," then stayed with the same dolphin for as long as possible. When our focal dolphin joined or left other dolphins, we followed it. At regular time intervals, we took stock of what he or she was doing, where, and with whom. This is not to say that we ignored the other dolphins in the group, but their importance was as context for the actions of our focal dolphin. Doing focal follows required intense concentration, particularly difficult when the dolphins were not doing much and the temptation to daydream was overwhelming. But we trained our eyes to see more and more of the subtleties of their behavior. At the end of a day we were rewarded with a great sense of a day in the life of the dolphin we had followed.

Andrew and I settle in to following Square. Talking into the tape recorder, I note the date, the time, whom we are following, and what she and the rest of her group are doing. For the rest of the duration of our follow, we will check in every five minutes with the same basic information and also describe in detail any interesting behaviors we see. Using our depth sounder, we will keep records of the water depth and temperature and approximately how far the dolphins have traveled. At the end of our follow, we will have systematic data on where Square went, with whom, what she did, how often, and for how long.

The group has settled down, spending long periods below the surface. We try to guess their trajectory and move along as slowly as possible. About every three minutes they float serenely up through the clear water, often as not right alongside our boat, break the surface to breathe, and snag, bobbing slightly in the glinting sun. We slow the boat and put it into neutral, our best approximation of a snag. I wish our motor were silent and hope the noise isn't too disturbing. They don't seem to mind.

After an hour it gets to be hard to concentrate. With the dolphins out of view much of the time, we spend considerable time staring at the hypnotic winking and wavering of hot sun on water. Andrew and I are tempted to chat, to fiddle with equipment, to daydream, to think about lunch. But even with a group like this, on a day like this, dolphins can disappear as if by magic. We try to concentrate.

Many young students have approached me and the other researchers, eager to gain experience watching dolphins. They are stricken with the romantic idea of watching dolphins, and they are certain it will be pulse-pounding excitement at every moment. Though there are plenty of thrills, I always feel that I must impress upon them that most of our time is spent like this: struggling to concentrate on the ups and downs of dorsal fins under a baking sun on a cramped little boat.

After what seems like a very long time, Square's group begins to rouse. Squarelet keeps turning in different directions, as if listening for some sound that will help her to decide which way to go. She can probably hear the sounds of other dolphins in the distance. Maybe she hears a group of youngsters playing, squealing raucously, and whistling out to the north of us, while another dolphin echolocates, a steady, creaking, rusty-hinge buzz of clicks, hunting out east of us. Maybe it is the deep, low "knocking" of males courting a female that catches her attention. Though we may have a hard time locating dolphins sometimes, I suspect that with their keen hearing they are usually well aware of who is where and what they are doing throughout much of Red Cliff Bay.

Back behind us now, just sixty feet and approaching, we see the source of Squarelet's interest. Two more dolphins, Joysfriend and Joy, are catching up. Square, Squarelet, Peglet, Tweedle, and Fatfin all snag as if waiting for the others to catch up. Just as they do, Squarelet flips up, bringing the whole front half of her body straight out of the water, wiggling as if she is being goosed from below. Which is probably exactly what is happening. Joy surfaces with her belly tilted toward Squarelet, hugging her and patting her sides comically with her pectoral fins.

The two roll and splash and chase at the surface, until suddenly Joy comes blasting through the water surface and soars into the air. There is a moment at the apex of her leap where she seems frozen in place, before gravity presides and pulls her back into the water with a tremendous splash. While she is airborne, I see her eye peeking at us. I try to imagine how that leap must feel, the sudden silence, the brilliant light and tremendous view, and then the tug of gravity and splashing reentry back into her familiar noisy, murky, weightless water world. She lands almost on top of Squarelet, who shoots out of her way, then swirls back around toward where Joy must be, hidden from our view below the surface. Peglet breaks away from Square and joins the fray. The youngsters, well rested, are ready to play.

Socializing takes many forms but pretty much always involves a lot of splashing and rolling around, bellies flashing to all points of the compass. Socializing dolphins don't travel in any consistent direction or maintain any consistent orientation of their bodies relative to the water or to one another. Sometimes all we can see is a lot of splashing and flashing. Other times we can get a sense of what goes on. Jawing, chasing, goosing, and sex play are common features of socializing groups.

Jawing is when one dolphin places its open jaws against the other, scraping or threatening to scrape with its teeth. The result in some cases are small, evenly spaced little tooth-rake scars on the skin of the recipient.

Goosing is just what it sounds like. One dolphin comes up from below, poking his or her rostrum into the genital area of another, who wriggles and flips over evasively. More often than not, there is a sexual component to these social interactions, particularly obvious among males, since they often sport bright pink erections, which they attempt to put into any orifice available.

Dolphins are notoriously "sexy." In fact, during Victorian times, the first attempts to keep dolphins in captivity were met with revulsion when males displayed their erections, using them to tow objects around with and masturbating in full view of visitors to the aquaria where they were housed. Dolphins are also notoriously playful. Behaviors we consider "sexual" are part of the normal social interactions typical of dolphins and may not always carry the same meanings as they do for humans. From a dolphin's perspective, just because his or her genitals are involved does not mean the behavior is sexual.

Square spends an hour or so tail-out-dive foraging while her daughters socialize with Joy and Joysfriend. We follow her from a distance, trying to stay out of her way, when I suddenly get a sinking feeling: Square should have surfaced by now, but I haven't seen her. This is just the sort of situation in which we are likely to lose her. Andrew and I both scan intently the full 360 degrees around us, looking for her dorsal fin.

Sploosh . . . a dolphin leaps out of the water, taking a deep breath as it flies through the air and reenters—a clean traveling leap. Then another. Abruptly we are surrounded by leaping dolphins, all heading as fast as they can go out to the north. Perhaps Square is among them. We gun the motor and head out into the midst. The drone of the motor puts a damper on conversation, but it's exciting to be zooming along with dolphins leaping alongside. At one point, as we are going full speed ahead, a dolphin veers over to the bow of our boat and streaks along in front of us, then leaps alongside before departing. It is Pointer, a young male.

I catch a glimpse of Fatfin leaping just ahead, which is encouraging. Square is probably nearby. After a few minutes the dolphins are no longer traveling. Instead there are many fins scattered around, all heading in different directions. Most are porpoising now—that is, they are surfacing rapidly and steeply, bringing most of their backs but not their bellies out of the water. They seem to have arrived at their destination.

One dolphin surfaces hard next to us with a fat silver fish lying cross-wise in its jaws.

The dolphins are doing what we call "leap feeding." All the dolphins in the area have somehow learned that there is a school of fish out here and have converged on this spot at top speed. Maybe the dolphin that discovered the fish had put out a call, encouraging others to come and help contain the school.

Leap feeding is usually indicative of feeding on schooling fish like mullet and bony herring. Now, since the dolphins are milling around in all directions, it looks as though the fish school has broken up into smaller pieces. One dolphin zooms past our boat in pursuit of a group of about five or six fish, then surfaces a moment later, gobbling one down.

It is virtually impossible to keep track of a particular dolphin during this activity. Surfacings are so brief and fast that it is difficult to identify a fin. But it can be done, so we struggle to do so, hoping to catch a glimpse of Square's fin in the confusion. When Peglet appears at our bow for a ride, we breathe a sigh of relief. Square is surely nearby.

Leap feeding is one of many different hunting techniques the Shark Bay dolphins employ. Other common strategies we discovered include tail-out-dive foraging, bottom grubbing, and snacking, not to mention the more unusual techniques like sponge carrying and ker-plunking.

A bottom-grubbing dolphin headstands, facedown and tail up, poking around among the seagrasses or into the sandy bottom after fish that hide there. It is usually done on an individual basis, so presumably the fish are solitary rather than in schools. When a fish is flushed and tries to flee to a new hiding place, the dolphin pursues it and often has to flush it out several more times before surfacing with the prize held in its jaws. Bottom grubbing, for some reason, is accompanied by a very distinctive sound, a descending *nyarrrrrr nyarrrrrr nyarrr*, that sounds very different from the typical echolocation sounds accompanying for-aging at other times.

Snacking is sort of the antithesis to bottom grubbing. A snacking dolphin turns its belly to the sky right at the water surface and pursues its prey, usually small garfish, echolocating intensely but also appar-ently using its binocular vision. Dolphins have binocular vision (over-lapping fields of vision, resulting in more accurate depth perception) just ahead and slightly below the body axis. Hence, it helps for a dol-phin to turn upside down in order to look at something just ahead and against the surface. The fish try their best to evade capture, twisting and turning, trying to outmaneuver the dolphin. Pursued from below, a doomed fish will ultimately find itself pinned against the surface. In a last desperate attempt to escape, snack fish often flip out of the water and sometimes land right in the dolphin's waiting jaws.

One hazard of snacking is the parasitism of seagulls and terns. They take advantage of the dolphins' efforts; after fluttering overhead until the moment is right, they drop down to seize the fish just before the dolphin can grab it. This aggravates the dolphins. Once I watched Snubnose grab a thieving gull by its leg, shake it, and heave the squawk-ing bird into the air.

After a time, the activity dies down and we find Square, Peglet, and Squarelet back together again. We resume our data collection. At least we know roughly how much time Square spent leap feeding. The family travels east and joins another group. Predictably, Fatfin and Tweedledee are there, along with Joysfriend and three young males, Lucky, Pointer, and Lodent. It's a petting session.

Joysfriend rolls underneath Lucky and turns upside down. Pressing the little bulge that protrudes from the point where the two sides of her tail flukes meet up against Lucky's outstretched pectoral fin, she waggles her tail end back and forth so his fin tip flicks back and forth across the spot: keel rubbing. A moment later the pair surfaces, and on the way back down, Joysfriend rolls underneath Lucky once again. This time she positions herself so that her belly is handy to Lucky's fin, but he doesn't seem to cooperate. She has to do all the work and rocks her entire body back and forth so that his fin, barely even out-stretched to accommodate her, moves back and forth across her belly. Another breath and Joysfriend passes over Lucky to Lodent. She posi-tions herself so that the tip of Lodent's pectoral fin is inserted into her genital slit and thrusts back and forth vigorously eight or ten times be-fore dismounting to take a breath.

Dolphin petting is a lot like "grooming" in primates. Chimpanzees, like other apes and monkeys, spend a lot of time picking through one another's fur with great care and dexterity, picking off tiny bits of flaked skin or ectoparasites. Grooming obviously feels good, like having some-one run fingers through your hair and stroke your back. Chimpanzees who are frequent grooming partners are either close friends and family or need grooming to help them calm down, maintain peace, or recon-cile their differences. Not too surprisingly, subordinate individuals are most likely to do the grooming, while dominant individuals get groomed.

As in chimpanzees, petting among dolphins is probably most com-mon among adult males, particularly during times of social tension, but it occurs between all age and sex combinations and in various contexts, both tense and relaxed. Joysfriend was an adolescent at the time of this observation. Such exuberant solicitations of petting by female dolphins, often with a strong sexual component, are most common at that age.

Joysfriend became more and more sensual and excited in her ap-proaches, while Lucky, Pointer, and Lodent seemed surprisingly lack-adaisical. They appeared to be tolerating her advances, but she had to do all the work. Perhaps they were just trying to play it cool. In any case, they cooperated with us and stayed close to the boat so that we could film their petting interaction with the videocamera.

Square and her daughters, along with Tweedledee and Fatfin, drift farther and farther from Joysfriend and the males, then begin to travel steadily back to the south in a tight knit group. Peglet rides in baby position, alongside and toward the rear of her mother. Andrew and I take advantage of the predictability of their travel to wolf down our peanut-butter-and-jelly sandwiches, apples, and thermos of tea, punctuated by our five-minute checkpoints, spoken stickily into the tape recorder.

When the time is right—we know just where our dolphins are and can easily keep track of them—we take a bathroom break. There is no way around it with long hours out on the water and no privacy or accommodations. It takes a certain amount of dexterity, particularly for women, and on a rough day, when the boat is pitching and rocking, it can be difficult to relax into the task.

It was only on the extraordinarily rare occasion that we were forced to stop everything and head back to shore for such necessities. One such occasion happened when I was out on the bay with Richard and his handsome young assistant, Eric. It was a chilly, windy, and exciting day. We had needed to completely focus our attentions on the dolphins for quite some time. I had hesitated to interrupt our follow, which had been fast paced and thoroughly engrossing, and had been trying to ignore my full bladder for much too long. But now my teeth were floating, and I insisted that we take a bathroom break. Richard and Eric complied, turning their backs to give me a modicum of privacy, but with the boat pitching and heaving, and the cold breeze against my rear end, there was no way I could relax enough to pee. After a minute or so Richard began making wisecracks and playing peek-a-boo. They were both giggling like a pair of twelve-year-old boys, and I had to give up.

Ten minutes later I simply had to try again. Repeat performance. By then I was beginning to shiver and feel woozy. I tried three or four times, with Richard and Eric more and more uproarious each time. Finally I had to demand an end to our follow and insist that we drive back to camp and anchor the boat while I used the facilities. By the time I made it to the bathroom I felt truly ill, and it took some time to recover my composure afterward. Richard reminded me of this particular incident numerous times over the years. Fieldwork forces the sharing of such intimacies, like it or not.

We spot some dolphins snagging up ahead. Square and her entourage head straight toward them. It is two groups of adult males: one group is Realnotch and Hi, and the other is Trips, Bite, and Cetus. The males seem to be asleep, snagged at the surface, eyes partially shut, barely a movement. They pay little heed to the females, who pass through their midst like ghosts through walls. Tweedledee is the only one who shows any sign of interest. She tilts in alongside Realnotch and rubs her entire length along his fin, then snags briefly in front of him before moving on. A greeting?

Dolphins are constantly encountering one another out in the bay, and their greetings run the gamut from nonexistent to ballistic and everything between. It is quite common that we see no sign whatsoever of acknowledgment as groups pass through each other. Tweedledee's behavior toward Realnotch, rubbing her length along his pectoral fin and then snagging by him, is a typical female-to-male greeting.

On a few occasions I have seen males greet each other in an especially curious fashion. While one male snags at the surface, another or even a few other males take turns passing perpendicular to him, brushing up under his chin and rolling belly up as they do so. When they surface, they usually snag alongside him for a moment. I have no idea what this means, but it strikes me as a way of paying respects.

Then there are those occasions when groups come together and there is complete pandemonium. Once we watched as the "B boys" (Beejay, Bumpus, Bam-Bam, Biff, Bo, and Biddle) charged headlong into a large group already containing most of the prime adult males in Red Cliff Bay. They all dove as the B boys arrived, leaving silence and a cloud of bubbles visible at the surface. But a hydrophone in the water relayed the screaming, hooting, hollering, growling, smacking, barking, screeching, and banging coming from down below. Then there was a sudden silence as the entire group broke the surface to breathe, all emerging from one central point, like the unfolding petals of a flower. After a few huffing breaths indicating their exertion, they immediately submerged again to resume the ruckus below. They did this flower-petal surfacing several times. I only wish we could have seen what was going on below the surface.

Greetings among females are usually more subdued. When they meet, females sometimes ignore each other, or they may travel alongside one another for a time or tilt their bellies toward each other briefly.

Square's group slows a bit as they come over the shallow bank. Squarelet headstands, bottom grubbing after some fish. Fatfin approaches her, and the two seem intent on something down below. In the clear water we can see to the bottom but can't see what they are after. Square goes down and comes back up with a little fish held delicately in her rostrum tip. She releases it just in front of Peglet. It looks as if she is trying to teach her how to hunt. Peglet spends several minutes chasing the tiny fish while her mother snags, turning this way and that, seeming unsure about what to do or where to go. Perhaps she is listening.

The sun is getting low in the sky, and we will need to head in soon. Peglet comes back over to the bow for a brief ride. She seems all wound up after the excitement of leap feeding, and I suppose it will take her a while to settle down. She approaches Square and pokes her rostrum in toward her mammary slits. For a moment Square slows just a bit, and Peglet remains there, nursing. They repeat this a few times, and I try to get a photograph.

As the sun sinks low, the sky and water take on the colors of a Maxfield Parrish painting: shades of iridescent blue touched with pastel yellows and pinks. At five-thirty we call it quits and pack up our equipment for the ride home. Everything goes back into covered buckets and waterproof cases for protection against the salty spray. We turn toward camp and pick up speed. Square and her group recede in the distance behind us as the motor drones and the boat picks up off the water, planing over the water surface, heading back to the terrestrial world. This is always a time to mull over the day's events.

Having spent most of the day with Square and her daughters, Fatfin and Tweedledee, we know what other dolphins they encountered, we watched their cycles of rest and activity. We know where they went. We let our own moods run parallel to theirs: restful and lazy at times, excited and alert at others. Because of it, I feel a renewed sense of connection with these dolphins. Also, I feel privileged that they allowed us to spend the day with them. I wonder what thoughts go through their minds about us as they listen to the sound of our motor fade into the distance.

As we approach shore, the dinnertime smell of frying fish wafts out to us, and the sounds of birds and voices of kids playing gradually become audible. After nudging the dinghy up onto the beach, I step out onto terra firma. My legs feel wobbly and unsure after so many hours on the water.

The Monkey Mia dolphins are all gone except for old Holeyfin, who paces offshore, coming in occasionally to approach people wading into the water. She approaches Andrew as we are unloading the boat. A small group of tourists descends upon her and therefore us, gathering around and trying to entice Holeyfin to come closer, riddling us with questions. Reentry into the world of humans and camp life can be disconcerting at times. We unpack all the equipment, anchor and tie up the dinghy, and haul everything back to our camp.

After dinner I wander down to the water's edge, bone tired and thinking of bed. Stars glimmer on the water, waves lap against my bare toes, the tide is rising fast. I wonder what Square and her girls are doing now.

Back at camp, I mention my thoughts about Square to Andrew. He jumps on it. It's calm and clear, a perfect opportunity to go out on the water and see what the dolphins do at night. I try to argue that I'm too exhausted, but Andrew convinces me that we should take advantage of the opportunity. We can always come right back in if we want to. We load up the boat with a minimum of gear; we won't be able to take photos, so there's no point in bringing a camera, and with such limited visibility, we will have to depend on our ears more than ever. We pack up the tape recorder and hydrophone.

At night the water is so black that it's more like oil. I don't like being unable to see what I am stepping on as we launch the boat. The metal dinghy is cold and clammy, and I feel chilled before we are ten feet from shore. As we push the nose of the boat out into the immense darkness, all I really want to do is to crawl into my nice warm sleeping bag in the safety of our tent. Especially because I am highly skeptical that we will be able to locate any dolphins whatsoever in the dark.

Traveling over the water at night is a strange sensation. Even in daylight, the landscape is wide open and featureless. At night there is nothing whatsoever on which to focus one's eyes, and we seem to be flying headlong into a void. I have this irrational anxiety that we are going to suddenly crash into a solid wall. After a few minutes, Andrew slows and cuts the motor. Now my anxiety turns to the motor. What if it won't start again and we are left drifting out here in the night? We assess where we are. A breeze is blowing, and the stars above twinkle and blink. The lights of the campground seem far away, but it is impossible to judge distances with no reference points. The lights beckon with the promise of warmth and comfort and safety.

It is so profoundly dark out here, we wouldn't be able to see a dolphin if it surfaced right alongside us. We put the hydrophone into the water and listen, tilting the dinghy to one side to muffle the splashing of water against the hull. I remember a much admired camp counselor who instructed me as a child on how to listen for animal sounds in the woods. He suggested closing my eyes and opening my mouth slightly. I try that now, straining to hear a dolphin breath.

We move on and listen at several different locations before we hear any dolphins. It sounds like two of them, and from the fast, hard sound of their breaths, they are probably hunting or traveling rapidly. Through the hydrophone I hear a sudden intense shower of clicks as the dolphins "look" our way with their echolocation. Then they seem to evaporate.

At night the dolphins must rely completely on their echolocation and hearing. Most sharks are more active at night, and it must be a dangerous time. Especially so during low tides, when all of the creatures in the bay are forced to share a few deepwater channels.

At another checkpoint, we hear the *sploosh-thud* sound of a dolphin repeatedly leaping clear of the water and slapping down hard on reentry. I have heard the same thing at night on a number of occasions previously. Maybe this is some sort of foraging activity that the dolphins do mostly at night; maybe the slaps help to herd some sort of fish they are hunting. Or perhaps the dolphins are trying to rid themselves of a skin irritant. We have seen dolphins do multiple high slapping breaches when a remora or fishing hook was stuck to them. Or maybe the slapping sounds are useful as signals to other dolphins.

After an hour or so, Andrew and I head back to camp. Just from listening to the breathing and surfacing patterns of the dolphins, we had at least been able to tell that they seem to do lots of things at night—resting, traveling, hunting. But it was impossible to see anything. Even when a dolphin had surfaced right alongside us, all we could discern was a dark shadow and the sparkle of moonlight on disturbed water. Any further inquiry into the nighttime activities of the dolphins would require night-vision scopes, which were well beyond our budget.

I was glad we had made the effort. There was a certain thrill about being "out there" at night. But I was even gladder to be back at camp, warm and dry and snug in my bed. I would have a hard time as a dolphin, swimming around day and night in the cold, wet, and wild ocean. I fell asleep recalling Peglet's face as she looked up at me from the bow of our boat earlier in the day.

DOLPHINS

AND SPONGES

L earning the basics of dolphin behavior, gaining familiarity with their daily activity patterns, and amassing information about where they go and what they do take many, many hours of following and watching. But a few of our important discoveries came about serendipitously. We weren't looking for them and didn't expect them. Sponge carrying was one such discovery.

It was September 7, 1984, during our first year of real research, with our new boat and the "onus upon us." We were just getting to know the dolphins and learning to watch them. Andrew and I had left a group of foraging dolphins and were taking a break, discussing what to do next, when we were interrupted by a dolphin surfacing one hundred feet south of us, a dolphin with a big rust-colored blob attached to its head. "What the hell?" we both exclaimed in unison, reaching for binoculars. We strained to get a better look, but the dolphin dove out of sight. As it brought its tail up and out of the water, we could see that one side of the tail flukes was missing. Guessing where it would come up next, we repositioned the boat accordingly. *Pfhooo.* The dolphin reappeared and moved slowly along, just below the surface, then came up for a breath, moved along again, and took another breath. Sure enough, there was the big, lumpy, reddish brown blob that seemed to be attached to the front of the dolphin's beak, extending back toward its face. Again the dolphin dove, bringing its maimed tail flukes out of the water as it went.

I pulled out the camera as we repositioned ourselves again. Up came the dolphin. This time we were lucky; it was just alongside the boat. And yes, the thing was still there. I eagerly snapped a few pictures, too hasty in my excitement to be sure whether they were even in focus.

"It looks like some kind of disease."

"Yup, like a huge tumor or growth of some kind."

I sat down hard on the hot aluminum bench, engulfed by a wave of pity. This poor dolphin was suffering from some horrid affliction and still having to forage or else die of starvation. I wondered if it was painful, if the other dolphins might avoid it. Sick, alone, suffering . . . My mind raced to think of some way we could help, but this was a wild dolphin and there was little we could do.

Back in 1982, one of the old fishermen had approached Richard and me on the beach to tell us that he had seen a dolphin with a big growth on its face. He had even mentioned that it was missing half its tail flukes, and recalling the conversation aloud to Andrew, I remembered how he had pointed out to the northeast and said, "It was out that way, in the channel past the first flats, about two minutes at full speed from here"—just about exactly where we were right now. Although we had heard lots of fish stories, some of which had been totally implausible, Richard and I had gone out to the spot he indicated to have a look around. We hadn't seen any dolphins at all, but we had seen a couple of dugongs. Skeptical, we had agreed that the old man must have confused one of the blunt-snouted dugongs for a dolphin with a growth on its face. Now I realized our mistake.

Then the dolphin surfaced again but this time there was no tumor. We raced to get closer before it disappeared again. Sure enough, although its dorsal fin lacked any obvious distinguishing nicks or scars, those unmistakable tail flukes flipped up into the air as it dove. Breathing a sigh of relief, we realized now that it wasn't a tumor at all. The dolphin was carrying something around. The story took a welcome turn from disease and suffering to an intriguing mystery.

We watched for several more hours, keeping track of the dolphin's dive times, number of breaths, and whether or not it had the blob on its face at each surfacing. We had been calling the dolphin "Growthface," but now we changed her name to "Halfluke." Early on during the follow, a youngster joined; Halfluke was a mother. The baby must have been somewhere nearby all along, but we had missed it at first. When mom dove, the baby would loiter around at the surface, sometimes coming over to our boat to ride at the bow and look us over. Halfluke came up with the blob on her face again repeatedly. In fact, she had it most of the time, surfacing without it only occasionally. But we still couldn't tell what it was or what she was doing with it.

We knew dolphins often carried things around, usually to play with. We had seen them carry bits of seaweed, plastic bags, and debris, draping their toy on a pectoral fin, flipping it off and catching it on the tail flukes, carrying it in the beak, keeping it just out of reach from any interested playmates. But this didn't look playful. The methodical way in which Halfluke surfaced, taking several breaths, traveling slowly but steadily along, and then diving with her flukes up looked quite serious. Her baby seemed bored with the whole affair. It seemed unlikely that we would solve this puzzle quickly.

Over the following months, we regularly came out to look for Halfluke and her baby. More often than not, we found her, and each time she was doing the same thing: carrying that thing around on her face, doing tail-out dives accompanied by her baby. Halfluke seemed to get used to having us around. Sometimes she would surface right alongside our boat, and once or twice she even came to ride at the bow. We were able to get a few good looks and some photographs of the blob. It was some sort of sponge, reddish brown, rough textured. She didn't seem to be holding it in her teeth: it was rather cone shaped, and she simply tucked her beak into the apex of the cone. Sometimes we could even see the holdfast sticking out in front. It also became clear that it was not always the same piece of sponge. Some were small, covering just the tip of her beak like a little cap. Others were big and floppy and partly obscured her face.

It turned out that Halfluke was not alone in this odd behavior. There were Spongemom, Bits, and Gumby as well. They all carried sponge, and they all had babies, so we knew they were all females. Collectively they became known to us as the sponge carriers. We watched them regularly, collecting whatever data we could that might help to explain why they carried sponges around. Usually they did the same old thing, which was really not so interesting to watch were it not for the possibility of an answer to this riddle.

But every now and then a sponge carrier would do something a little different, like come up without the sponge and remain in place at the surface rather than moving along. It was as if she had left the sponge below and were trying to stay in place in order to retrieve it next time she dove. Once in a while a sponge carrier would suddenly leap or porpoise once or twice, a burst of excitement that contrasted with their otherwise plodding ways. Usually when they did one of these excited surfacings, they were not carrying the sponge, and on a few occasions we saw them chewing on something that looked like a fish just afterward.

On rare occasions, the sponge carrier we were watching might join up with another sponge carrier, usually just for a short time. There they would be, swimming along together, each with her sponge. Camaraderie born of a shared bizarre habit. Members of some strange dolphin cult. They certainly didn't associate much with any of the other dolphins in the area.

Why carry around a sponge? We entertained all sorts of possibilities. Could they be eating them? Could they be using them to lure some kind of prey? To conceal themselves while hunting? Maybe the sponges served some sort of social function? One good-humored suggestion was that these were the dolphin untouchables, whose job it was to scrub the bottom of the bay. Without getting underwater to see what they were doing with the sponges down below, we could only guess.

I tried getting into the water with Halfluke on a few occasions. It was nerve-racking. The channel was deep and murky and probably full of sharks. Each time, I would catch a brief, hazy glimpse of a dolphin at some distance, heading rapidly away from me. They were accustomed to people in boats, but not in the water with them. I imagine they probably thought humans consisted of head, arms, chest, and the hull of a boat, with a propellor attached to their rear ends. It was apparently alarming for them to see all those legs and arms flailing around in the water. In any case, we would have to come up with some other plan to see what they were doing with the sponges.

Once while we were searching for a sponge carrier to watch, we were surprised to come across Puck. Dear, familiar Puck, whom we thought we knew so well, thoroughly surprised us by carrying a sponge around! We had never seen any dolphins other than "the sponge carriers" carrying sponges, yet here she was. She acted as though she were an old hand, tail-out diving and surfacing with the sponge placed over her beak just so. She even swam right up to our boat with the sponge placed delicately over her rostrum. Fred, an adolescent male, followed her around attentively the whole time, but he never picked up his own sponge. Puck showed us that other dolphins do carry sponges, at least sometimes. Certainly they seem to know how to do it.

Meanwhile, Richard Wrangham, who studied chimpanzees in Africa, had been telling us about how the chimps self-medicated themselves using certain types of plants. They would consume *Aspilia* leaves just after getting out of their nests in the morning, selecting a few, fresh young leaves and swallowing them without chewing. The leaves of the same plant were used by local humans to relieve gastrointestinal problems and contained an unusual chemical that made them effective. Elsewhere, chimpanzees ate other plants in ways, and at times, that also suggested they were medicating themselves. This discovery made a big splash among zoologists, and suddenly people were seeing all sorts of examples of medicinal plant use. Using certain plants that were not otherwise a normal part of their diet, elephants were thought to be inducing labor, howler monkeys controlling the sex of their offspring, and black bears protecting themselves from insects. At UCSC I had come into contact with Phil Cruz, who studied the chemical compounds found in sponges. Sponges are notorious for producing odd chemicals, many of which have valuable medicinal properties. In fact, the pharmaceutical industry provides major funding for research on sponge chemistry. Were the dolphins medicating themselves with some compound in the sponge?

All of the sponges carried by dolphins seemed to be the same type, just different sizes and with some variation in color. If we could get our hands on a sample, perhaps we could have its chemical composition tested. We searched for hours over the shallow banks in clear water, scanning the bottom for sponges, but we didn't find a single one. The sponges just didn't seem to grow in shallows. We would have to dive for them right in the channel where we saw sponge carriers most often, but we didn't have the equipment for the job.

Eventually a dive team came to visit Monkey Mia and gladly accepted the challenge. They pulled up a great collection of odd sponges and soft corals, almost all of which we had never seen on the adjacent shallow flats. We took photographs of each and then shipped the one that looked like dolphin's sponge to John Hooper, a sponge expert in Queensland. He wrote back to tell us that the sponge was *Echinodictyum mesenterinium,* a mouthful to pronounce but apparently widespread along the west and north coast of Australia. John also told us that a dolphin researcher working off the coast of Northern Territories had told him about dolphins with reddish globular sponge-like things on their beaks. Perhaps dolphins carried sponges even outside Shark Bay.

We had a sample of the sponge tested preliminarily for unusual chemicals, but there didn't appear to be anything out of the ordinary. John Hooper tested them to see if they were toxic to common bacteria. They were not. Although it is possible that there is some hidden chemistry yet to be discovered, the self-medication hypothesis was losing credibility. Besides, the sponge carriers didn't act as though they were sick. They all seemed quite healthy, and all had successfully reared babies. By that time, we had watched them for several years, and they continued to carry sponges most of the time. If they were motivated by some illness, then it would have to be a chronic, female-specific illness.

The failure of the self-medication hypothesis left us with only one hypothesis that fit with all of our observations. The dolphins must be using the sponges as a tool to aid in some sort of foraging. Their tail-out dives indicated that they were heading straight to the bottom when they went down with the sponges, so whatever they were doing, they were doing it down there. The amount of time that sponge carriers spent going about their strange business was similar to what normal females spent foraging. They had to be eating during that time, and we did see them occasionally surfacing with prey in their jaws, chewing.

The divers who collected sponges for us described the bottom as "sandy, with occasional rock outcrops and swept by a strong current." They also reported seeing a few scorpionfish (also called lionfish), fanciful-looking creatures covered with spines and frills, brilliantly patterned in black and white stripes and spots. The spines contain a powerful toxin, and if the fish is alarmed, it erects the poisonous spines. Touch the spine and you are likely to regret it for a very long time. A biologist who researched scorpionfish once visited with us at Monkey Mia. He had been handling them in aquaria in his lab for years without incident, but ironically, while he was on vacation snorkeling in the Red Sea, he had accidentally put his hand against a scorpionfish and was stung. That was two or three years before we met him. He said the pain had been excruciating for well over a year. Much of his hand looked as though it had been severely burned, the tissue having simply melted away. This kind of damage to a dolphin's face would be devastating.

There are lots of other stinging creatures to watch out for in Shark Bay besides scorpionfish. Startle a stingray, for example, and it will whip its barbed tail at you. If you are struck, the barb can become lodged in your flesh. We had heard reports of dolphins dying from these barbs, which can eventually work their way inward, lodging in vital organs and introducing infections. Stonefish also are relatively common in Shark Bay. They hide among rocks on the bottom and, like scorpionfish, have toxin-laden spines.

All of these nasty creatures could be hazardous to a foraging dolphin, and some are particularly common in the channel. The dolphins are most likely using their sponges to shield themselves from the spines, stingers, and barbs of creatures they encounter. They could also be shielding themselves from abrasion. Dolphins sometimes poke their beaks into the sand and bottom debris after burrowing fish. Using a sponge might help them to avoid getting scratched and cut by small sharp bits of shell and stone.

There was a time when tool use was considered to be one of the "hallmarks of the human species." Only humans, with their advanced intelligence and their opposable thumbs and manual dexterity, could make and use tools, the reasoning went. But then Jane Goodall reported tool use in chimpanzees. Chimps use long, straight twigs to poke into termite mounds. The termites, defending their mound from intrusion, clamp onto the twig, and the chimp can then pull out the twig and eat the attached termites. Chimps are selective in choosing twigs that are just the right dimensions and sometimes break the twigs or peel the bark off to make them more suitable for the purpose. Chimps in West Africa also use "hammer and anvil" stones to pound open hard nuts. In fact, quite a few birds and mammals are tool users, and some can even make their tools. The New Caledonian crow, for example, makes quite a sophisticated tool by stripping pandanus leaves to obtain just the spine of the leaf. It then bends the tip of the spine to make a barb, which it uses to jab inside holes in tree bark to spear grubs.

Prior to our discovery of sponge carrying, dolphins in captivity had been seen to use tools in various innovative ways. For example, at the Port Elizabeth aquarium in South Africa, a dolphin took to using pieces of broken tile to scrape against the tank walls, dislodging seaweed that he then ate. Later, a second dolphin began to do the same. In another report, two dolphins were observed harassing a moray eel in their tank. The eel hid in a crevice. After many attempts to get it out, one dolphin sought out a scorpionfish. Carrying the fish carefully by the belly to avoid its spines, the dolphin used the poisonous scorpionfish to poke at the eel. The poor eel was forced to abandon the crevice and was then caught by the dolphin.

These examples from captivity, interesting though they are, all involved innovative behaviors that occurred on very rare occasions, by single individuals. Our discovery of sponge carrying captured the attention of many people not only because it involved wild dolphins using tools in their natural environment to enhance their survival, but also because sponge carrying is a consistent behavior, probably learned and passed on from generation to generation: a cultural tradition of tool use.

Many years have passed since we first discovered sponge carrying, and still, Halfluke, Spongemom, Bitfluke, and Gumby can all be found with their sponges, out in the usual place. Halfluke's baby, whom we called Demi, grew up to become a sponge carrier herself. Once in a while we see other dolphins carrying sponges, but it is always brief. As we extended our range to explore new areas of Shark Bay, a few new dolphins have been seen with sponges. So now we know it happens elsewhere in Shark Bay, not just in the nearby channel we refer to as "spongeland."

Though the dolphins probably use their sponges in some way as shields while they forage, we still don't know for sure, and we certainly don't know exactly how they use them. Nor do we know why only a few females seem to specialize in sponge carrying. It will take years of effort, some innovative techniques for watching the dolphins underwater, and a ton of good luck to make the observations needed to completely solve this mystery.

Guichenault Point, near our campsite. Steep cliffs and red dirt are typical of Shark Bay. (Author)

Dolphins cruise over shallow sand and seagrass-covered flats typical of much of Shark Bay. Water clarity depends on the weather, but after a period of calm winds, it is easy to observe dolphins through the water surface. (Andrew Richards)

Snubnose earned his name from his upturned rostrum. (Author)

People can learn to tell dolphins apart by the natural markings on their dorsal fins, but dolphin faces are also unique. Bibi always looked a little crazy. (Author)

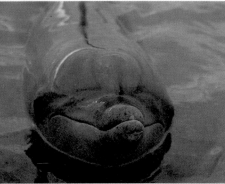

Holly bringing a piece of seagrass to play. (Author)

Puck, daughter of Crookedfin. (Author)

Nicky, daughter of Holeyfin. (ANDREW RICHARDS)

Crookedfin. A fine stream of bubbles rising from her blowhole indicates she is whistling. (ANDREW RICHARDS)

Holeyfin, the grand old lady of Monkey Mia. (AUTHOR)

Different dive types are typical of different kinds of foraging and travel. A peduncle dive indicates the dolphin is going down at a relatively shallow angle. (Author)

Smoky, son of Yogi, poking his rostrum into weeds on the bottom while looking for fish. We called this type of foraging "bottom grubbing." (Author)

A tail-out dive indicates the dolphin is heading straight down toward the bottom. (Andrew Richards)

Dolphins in Shark Bay leap and "porpoise" (surface rapidly with the back exposed but the belly still in water) while foraging after schooling fish. Dolphins are attracted to leap and porpoise feeding from all around. (AUTHOR)

Holeyfin and her daughter, Holly, petting. (Barbara Smuts)

Holeyfin with her newborn calf, Nova. Note the scars on Holeyfin's back from the sunburn incident. Nova still has "birth rings," or fetal folds surrounding his body.

Dolphins pet and rub. (AUTHOR)

When dolphins rolled belly up, we could see both their genitals (and therefore determine their sex) and their belly speckles, an indicator of age. This dolphin is Bibi, a young adult male. (ANDREW RICHARDS)

Shark Bay is a place of spectacular sunsets — complete with leaping dolphins. (Author)

Playing with wild dolphins in their own element is not only fun but also provides a good opportunity to learn about the dolphins' manners, expressions, moods, and creativity. (Andrew Richards)

THE

BIG BANG

Over the years, I grew accustomed to traveling back and forth to Australia. Even the interminably long airplane ride became second nature. I knew where to go for a peaceful nap in the gardens at the Hawaii airport, where we always stopped to refuel in the middle of the night. I became familiar with the rituals and hassles of going through customs on both ends of my journey and learned a few tricks for coping with serious jet lag. Because Western Australia is about twelve hours ahead of the United States, the lag amounts to a complete day-night reversal.

Forcing myself to do without sleep during the first day or two after arriving in Australia helped me to adapt more quickly, but coming back the other way was somehow always more difficult. For weeks after returning to the United States, I felt groggy during the day and wide awake during the night.

Most of the time, I traveled with Andrew or a research assistant (we had no problem finding capable volunteers, usually undergraduate students seeking some experience). After a few days in Perth gathering together the last bits of equipment and some special groceries not available in Shark Bay, we would make the trip, one way or another, up to Monkey Mia. So the entire journey, from home (California in those days) to Monkey Mia, could take a week or so. Cruising up over the last rise in the road, usually at dusk, we were met by a gorgeous view of the bay, with the campground lights glowing against the shore.

In 1985 I made the trip on my own, and I remember arriving back at Monkey Mia most vividly. Stepping out of the car after the long, hot, ten-hour drive from Perth, a welcome swallow swooped down over my head, snapping its bill and sending a little white turd directly onto my shoulder. Welcome indeed. I was curious to see whether the dolphins would remember me. I hoped they would. In my fantasies, they were as thrilled to see me as I was to see them, leaping and cavorting to show their pleasure and lingering long against my outstretched hand.

I waded into the water where Nicky, Holeyfin, and Holly were consorting with a small crowd of tourists. Nicky approached me, and I searched for a glimmer of recognition. She came toward me, opened her mouth in a typical begging posture, then after a moment turned and swam away just as she does with hundreds of other people every day. No sign of recognition whatsoever. Dolphins don't have the pliable, intricately muscled and expressive faces that humans and our primate relatives have, and as a result, their limited facial expressions reveal little emotion. But there was absolutely nothing anywhere in Nicky's behavior to suggest that she recognized me.

Why should she? I thought. They have all probably interacted with so many different people since my last visit that my face (or whatever features they used to identify people) was lost in a sea of anonymity, a member of the general category human: one who gives us fish handouts, plays silly games with seaweed, and is always trying to touch us with those strange wiggly-grabby appendages.

Then Holly approached me. Just before I had left Monkey Mia the previous fall, Holly and I had developed a private little game. I would tap her gently with my fingers near her blowhole (something that is now forbidden, and with good reason). At the same time, I would make raspberry sounds through my pursed lips. She would respond by making raspberries back to me through her pursed blowhole. I leaned over and tapped my fingers alongside her blowhole. She hesitated, turned abruptly and looked me over, then enthusiastically responded with raspberries that resounded up and down the beach. She remembered our little game. She remembered me.

Holly's welcome was a delightful and much needed salve for my heavy heart. A month earlier, my father had died suddenly. Andrew and I had just returned to Perth after several months traveling around in Southeast Asia. We were just about to head back up to Monkey Mia when the phone call came. My mother's voice was raspy, and I could tell immediately that something was dreadfully wrong. "Rachel, I have some bad news. . . ." She sobbed a few times and then handed the phone to a friend who was with her, and he informed me that my dad was gone. I hung up the phone and collapsed on the floor of Bert and Barbara Main's living room. My father was only sixty-two years old. He always seemed healthy and, at least in my mind, invincible. He was never an outgoing, chatty man, but through his actions he had set an example for me, and I idolized him. I couldn't fathom the idea that he was gone.

My mother would need me, so the next morning, in a haze of grief and disbelief, I boarded a flight back to the States, to Long Island, to my family, to the house where we grew up, now with an empty spot at the dining room table where my father had always sat. His things were still around the house, dirty laundry still in the hamper, coffee cups on his desk. But he was truly gone.

The morning of my dad's funeral, I cajoled my mother into taking a short walk. We wandered down to Sunwood, an old estate built on the Long Island Sound beachfront. Sunwood, just down the street from our house, had been my childhood hangout. I had played on the beach, swum in the protected waters, wandered in the overgrown gardens, attended university events in the old mansion, ridden my horse across the sandflats. Now, with my mother and our grief, we strolled through the rhododendron-lined avenue and down to the beach, where waves calmly lapped against the rocky shore, gulls broke open mussels against the rocks, and old squaw ducks chanted in the near-shore waters.

My father, an avid bird-watcher since his childhood, a biologist by training, a conservationist in action, had spent much of his life communing with nature and fighting to protect it. In the early 1960s he and some of his friends founded the Environmental Defense Fund, an organization that has now grown into one of the largest environmental advocacy groups in the world. He was Jewish by heritage, my mother Catholic, and their marriage had caused friction on both sides of the family. But they had come to recognize a different religion: that of the natural world. As children, my brothers and I were not indoctrinated into any particular religious sect but rather taught to understand and appreciate the "great web of life." The teachings of the natural world seemed so much more tangible to me than anything I read about Christianity or Judaism. I was not expected to accept anything purely on faith. I could see with my own eyes the interconnections that make up the ecosystem. I was not expected to accept that some man up in the sky called God had created everything (and my kind as an afterthought, from the rib of his special creation, man). There was an explanation for how things came to be that fit with what I saw around me, with what we know about the history of life, and with common sense.

I knew my father would decay and his body would provide organic matter that would recycle into the great chain of life. Although he might transmogrify from human to bacteria to fungi to tree to whale to bird, he would remain part of that grand, fantastic whole. Walking along the beach at Sunwood with my mother, I imagined his spirit flowing around us, becoming part of the trees, the air, the ocean, the ducks, ourselves, the seagulls and mussels and sand and wind. There was a tangible sensation as our grief lifted and dissipated. My mother and I returned to the world of funeral arrangements and sympathetic visitors refreshed and consoled by this reminder of the fact that we are part of something much bigger than ourselves and our ephemeral human lifetimes.

After seeing my mother through the first weeks of her transition to life without my father, I returned to Australia. Leaving her and my childhood home behind once again, I felt sad and also optimistic. All those memories I carried from my childhood here were still so much a part of who I had become. But my beach wasn't Sunwood anymore, it was Monkey Mia, on the other side of the planet. And rather than old squaws, there were silver gulls and black-and-white pelicans, pied cormorants, and Holly blowing raspberries.

My father had been proud of me for following my heart and my mind to Monkey Mia, and I knew he would be pleased that I was back at it. So I arrived back at Monkey Mia both saddened that I would not be able to share my experiences with him and more eager and determined than ever. Andrew had been at Monkey Mia for a month already. Camp was all set up, the boat up and running, and there was news of some new dolphin babies, some females who looked pregnant, and the usual campground politics.

Holly had grown noticeably during our absence, and was more people centered than ever. She seemed to take great pleasure in playing little games with people, staying just out of reach, teasingly, then rushing in and pushing herself into someone's arms, playing give-and-take with bits of seagrass. Puck and Holly together devised a new game. They would graciously accept every stinky, slimy, or half-frozen mulie that was offered to them, but they wouldn't swallow any of them. Instead they collected great mouthfuls, six or eight fish at a time. They would then swim up to someone and spit them all out in a slimy mass of broken fish parts, scales, and guts, whereupon they would wait patiently for their human playmate to pick each one up and offer it back. Yet another variation on the give-and-take theme.

Puck and Holly seemed to feed off each other's enthusiasm for this game. When one started, the other followed suit. They watched each other out of the corners of their eyes as they made the rounds, mulie tails protruding from between their teeth, and jealously guarded their collection of fish to ensure that the other didn't get any.

Once Holly collected about eight fish in her mouth and deposited them in front of me. She then wandered off, approaching other people for more handouts, playing games with seagrass, and poking around. Now and then she glanced over at me, apparently checking to make sure I was still taking care of "her" mulies. When Puck approached me, Holly rushed over, apparently concerned that I might allow Puck to take some of her fish. She pushed Puck away and took the mulies all back, held them in her mouth for a while, then spat them out in front of me again. By now the fish were beginning to disintegrate and break into pieces. They were truly disgusting, but Holly seemed quite attached. I had apparently been selected to baby-sit her fish for her, and I happily obliged. I was not the only one to play these games with Puck and Holly. The number of tourists had increased noticeably since our previous visit.

Recognizing the growing tourist attraction that the dolphins were becoming, the local shire government, the Western Australia Tourism Commission, and the Department of Conservation and Land Management (roughly equivalent to our Department of Fisheries and Wildlife) agreed to cooperate in developing some sort of plan for managing the dolphins' interactions with tourists. This was essential. More and more people were flocking to Monkey Mia every day as word got around about the friendly dolphins. Where there used to be a handful of people at the water's edge, now there might be upward of a hundred or more. They were rolling in by the busload.

A few local Denham residents were hired to work as rangers, talking to the visitors, instructing them on how and how not to interact with the dolphins, and maintaining some basic records. It was agreed that the feeding of mulies had to stop and some controls put in place on the amount of fish fed to the Monkey Mia dolphins. They were getting fat and spending virtually all of their time in the shallows begging for handouts rather than hunting on their own.

Fresh, locally caught fish were the ideal alternative, so Andrew and I decided that we would chip in and try to catch some ourselves. We borrowed a net from one of the old-timers and took up fishing. Almost every evening we set off down the coast to an area that seemed to have potential. After staking one end of the net to the beach, we backed off the beach, letting the net unravel, sliding over the gunwales and into the water as we went, then anchored the tail end. As the sun set, we made ourselves comfortable on the beach, often with a small fire, and waited. Occasionally we waded into the water and felt the net. A good strong vibration was a sign of fish. On a good night, we would hear flipping and splashing and see the whole net bow as a school of fish pushed into it.

Pulling in the net was always an adventure. We were excited by the prospect of a good catch, but often enough we caught things we really didn't want to catch. Our very first night out, for example, we hauled aboard a large ray, badly entangled, then a small shark that had rolled over and over and over, ripping holes in the net and making a huge tangle. Then came the coup de grâce: a poisonous sea snake. We felt bad to have bothered the poor innocent creature, but we also feared getting bitten, so after much deliberation, with no other safe alternative, we bludgeoned it to death with our anchor and returned home feeling miserable, disgusted, and discouraged. But other nights were more successful. One night we pulled in a net packed solid with over two hundred fat bony herring, a dolphin favorite. The next day, with the help of several visitors, we fed fish after fish after fish to Nicky, Puck, and Holeyfin. It was good to see them eating something of better quality than the bait-fish mulies.

With all our fishy bounty, we decided, just for the sake of curiosity, to try offering some fresh dead fish to some of the offshore dolphins. Would they accept them, or would the fact that the fish were dead and came from human hands prevent them from eating? Was it their disdain for dead fish that kept others from exploiting the generosity of tourists in the Monkey Mia shallows? Over the next few days we carried a few fish with us in the boat. Wave eagerly accepted our offering and gobbled it down, then circled the boat for more. So did Shave, though he poked at the first fish we offered to him a few times before eating it.

One of the most interesting responses involved an older male, Steps. He was traveling with five other dolphins. When we offered him a fish, he took it immediately but didn't eat it. Instead he rolled onto his side, holding the fish crosswise in his jaw so that the fish's tail end and

Steps's pectoral fin stuck up into the air comically. He was whistling very loudly. Steps's partner, Midpoint, approached and began to stroke him on the side with his pectoral fin. Meanwhile, the four other dolphins in the group gathered around Steps in a semicircle, pointed toward him. He remained on his side for another thirty seconds or so and then dove with the fish. The other dolphins also dove with him. When they resurfaced, the fish was gone and they were back to traveling in a rank abreast.

Was Steps showing off his prize to the others? Why did Midpoint stroke Steps? Was he acknowledging Steps's bravery or encouraging him to share the prize? What was the meaning of the whistles? The whole interaction was fascinating but indecipherable to me.

Still, we had answered one question. Some dolphins had refused our offerings, but in general the offshore dolphins were not above taking fish offerings (though they didn't take them from hand). Eating dead fish didn't pose a problem. Apparently they had other reasons for not adopting the practice of coming to the beach at Monkey Mia. Perhaps the habituated dolphins somehow inhibited others from taking advantage of "their" resource? Or maybe they were willing to accept our handouts but not motivated enough to overcome their inhibitions about approaching people.

There were some exceptions. One old dolphin, Spike, began coming in to Monkey Mia and taking fish handouts around the time his health went into decline. Apparently taking fish from people was something of a last resort for him. We fed him by hand for a month or so, but his condition worsened and he eventually died.

We didn't want the offshore dolphins to learn to expect fish handouts from us. In fact, we wanted to cause as little disruption to their natural behavior as possible, so we limited these fish-offering experiments to just a few occasions.

Of even greater interest to us were the diverse ways that the dolphins went about catching fish for themselves. It seemed that there were as many different strategies as there were types of fish. One that especially caught our eyes (and ears) during our visit to Monkey Mia in 1985 was "bony banging," another discovery that came as a complete surprise.

*B*ang! I was "plugged in," headphones on, tape recorder rolling, hydrophone (underwater microphone) draped over the side of the boat, when I was nearly knocked overboard by a painfully loud sound. The needle on the tape recorder meter spiked into the red "overload" zone. The dolphins were feeding on a school of fish. Besides the crack, there was intense buzzing and clicking of all eight dolphins echolocating. Then another *bang!* Just at the same moment, a silvery fish hurtled a couple of feet into the air and then landed on the water surface. Instead of swimming away, it lay on its side, as if stunned. A dolphin snapped it up and gobbled it down. *Ba-bang!* Two little splashes of water happened simultaneously with these bangs, and from one, a fish came flying out. Just as before, it landed on the surface, stunned, then was gobbled up by a dolphin.

Since we had first started listening and recording the dolphins' sounds, I had heard new and interesting sounds virtually every day, but this was something that commanded immediate attention. How were they making these bangs, and what did they have to do with the fish flinging out of the water?

Back in Santa Cruz, Ken Norris had been expounding his latest pet theory: that dolphins are capable of producing sounds so loud that the sound itself could stun a fish. Ken and his colleague, Bertol Mohl, had just published a paper detailing all the evidence in support of their "prey-stunning hypothesis." Sound is composed of pressure waves, and the idea was that these pressure waves could pack enough punch to actually damage the fish, perhaps by overloading their receptors or even by damaging tissues.

There is no way a human could yell loudly enough to stun a flea. But the underwater world of sound and the ways in which dolphins make use of it bear little resemblance to anything we are familiar with. Most mammals have a larynx with vibrating vocal chords, but dolphins are not like most mammals. In fact, nobody has really figured out exactly how dolphins actually make their sounds. But we do know that they are acoustic creatures par excellence. Many species have excellent vision, too, but in murky water and at night, vision is simply not enough.

Because water is so much more dense than air, sound waves travel faster in water than in air. That point was brought home to me many times when, through my hydrophone, I heard a distant outboard motor start up, followed a second or two later by the same sound in air. The sound waves took that much longer to reach my in-air ears.

The difference in density between seawater and air is so great that when a sound pressure wave moving through seawater hits an air interface, like the surface, or the air-filled swim bladder of a fish, or the lungs of a dolphin, it can barely make the jump into the new medium. Instead the air acts as a mirror, reflecting most of the energy back into the water. If you are standing right next to a vocalizing dolphin, sounds that are loud and clear through your hydrophone will be barely audible in air.

Conveniently, fat (as well as other body tissues) has about the same density as seawater, so fat is an excellent conductor of sound in water. That means sound waves can travel from water into fat without losing much energy. Dolphins have taken special advantage of the sound-conducting properties of fat. The bulbous forehead of a dolphin is not just an adornment to make him look more adorable; it is actually a blob of fat, called "the melon." Although we still don't know exactly how dolphins make their sounds, we do know that they talk through their foreheads. By squeezing air around in a complex of air sacs and valves inside their heads, just below the blowhole, they create vibrations. The melon acts sort of like a lens, conducting the vibrations out into the surrounding water, just in front of the dolphin's forehead.

Fat is also used to conduct sound energy into the ear. It certainly wouldn't work for a dolphin to have big flappy external ears like those of many terrestrial creatures. They would get full of water and cause serious hydrodynamic drag. Instead dolphins have pads of fat located on their lower jaws that conduct sound vibrations through the thinned-out jawbone and on to the dolphin's inner ear bones. No air in the way, no floppy appendages, just liberal quantities of carefully deposited fat.

With all this unique anatomy, dolphins can produce and hear an incredible array of sounds. Tests have revealed that they can hear frequencies well over 100 kilohertz. A few lucky humans can hear up to 20 kilohertz at best. By producing a train of "clicks" and listening to the returning echoes, dolphins can essentially "see with sound." They can locate and track prey, discriminate between objects made of different materials even when they *look* exactly the same, and discern subtle differences in texture, size, and orientation. Some evidence suggests that dolphins can even look inside one another: by perceiving differences in the comportment of bone, tissue, and air sacs, they can detect something about the mental and physical states of their associates.

It is quite possible, given the different and very sophisticated ways in which they produce and hear sounds, that dolphins can, in fact, emit sounds that would be loud enough to stun a fish. Exposure to very loud sounds is known to have profoundly damaging effects on animals, including humans. The tiny "pistol shrimp" has a specially designed claw that it uses to shoot its prey. The shrimp can create a tiny vacuum space in its claw. Then, by releasing a pressure valve, the vacuum is snapped shut, creating a potent little bang. When a tiny fish passes nearby, the shrimp extends its claw toward its prey and shoots. The fish fumbles and gasps, too stunned to escape. Could dolphins have a similar effect on their prey?

The prey-stunning hypothesis made sense of a few odd observations. For example, some sperm whales have been known to survive, obviously eating well in spite of having broken and nonfunctioning jaws. How could they possibly catch their prey in that condition? Dolphins have been trained to emit very loud clicks within the range known to be lethal to some fish. But do they actually do so under normal conditions? In spite of all the circumstantial evidence suggesting that dolphins might have a built-in stun gun, nobody had ever observed prey stunning in action.

Could this be it? Were the Shark Bay dolphins actually zapping fish with so much force that they were sometimes blasted right out of the water? That seemed unlikely, but maybe the fish were leaping out of the water on their own, in a last desperate attempt to escape from being zapped. We tried to get a better look, moving right into the midst of the feeding activity.

The dark shadow of the fish school loomed just beneath us. Occasionally part of it rose toward the surface, as dozens of fish raced upward. As they approached the surface, they veered off in all directions, away from the dolphin in pursuit. The dolphins were echolocating so intensely, it was unpleasant to listen to, like a swarm of very angry, very loud bees. An all-encompassing *bzzzzzzzzzzzzzzzz* that made my skull vibrate and through which it was difficult to think.

Some dolphins hung around on the periphery of the fish school, apparently taking a break, maybe helping to keep the fish school from breaking up. We heard several bangs but saw nothing. Then a dolphin rose toward the surface, fish fleeing directly ahead of it, and suddenly swung around, as if abruptly changing direction. *Bang!* There was a splash of water, but no fish flying through the air. Before we could make sense of what we were seeing, the dolphin and the fish were gone again.

We watched all that day as the dolphins pursued the fish in clear water, just eight to ten feet deep. Sometimes they were all in one area, keeping the entire fish school balled together by swimming around and under it, echolocating. On a few occasions small parts of the school, a dozen or so fish, would break away, with a dolphin in hot pursuit. Once or twice the entire activity split into a few centers, as though the fish school had broken into pieces. A core group of dolphins remained busy much of the day, while some others came and went or hung around socializing on the periphery.

Since the dolphins were hunting bony herring, we called this type of foraging "bony banging" and made a point of trying to watch and record as often as possible. Sometimes, especially in shallow water, fish would come flying out of the water in conjunction with the bangs. In deeper water, we heard the bangs but usually saw nothing other than an occasional dolphin surfacing with a fish in its jaws just afterward. Once in a while we saw a dolphin swing its body around suddenly just at the bang. Did the swinging have something to do with producing the bang? Or were they simply swinging into position to grab the stunned fish? It always happened so fast, and when it was close enough to the surface for us to see, all the splashing usually obscured our view. We just couldn't make out the details.

We eagerly told Ken Norris all about our observations of bony banging. He was very excited by the possibility of proof for the prey-stunning hypothesis, but Randy Wells was skeptical. Randy had watched dolphins off the coast of Florida for many years, and he described to us a behavior he called "tail whacking." The Florida dolphins would chase mullet and at the last moment turn sharply and swat them into the air with their tail flukes. He thought this was what our dolphins were probably doing.

We just weren't sure. We had seen our dolphins swing around abruptly at the bangs, but Randy's description didn't really match what we had seen all that well. Ken thought it unlikely that the bangs were made by the dolphins actually hitting fish with their tail flukes. He reasoned that the tail flukes sweeping through water would push along a wall of water that would prevent the fish from actually coming into contact with the tail. Maybe, Ken suggested, the bangs were made by the dolphins vocally but didn't actually stun the fish so much as disorient them enough to make them easier to grab.

Another possible explanation for the bangs was cavitation, which happens when a force is applied that is powerful enough to actually split water molecules apart, creating a vacuum space that collapses with an audible bang. Cavitation is mostly the concern of navy engineers who design ship propellers, because they want propellers that won't cavitate. But marine mammal biologists noticed that when whales and dolphins accelerate rapidly, by sweeping their tail flukes up and down forcefully, they can actually create a cavitation bang. Maybe the dolphins were swinging their tails at the fish to disrupt the integrity of their school and, in the process, creating a cavitation bang.

With Ken's help, we purchased a videocamera and an underwater housing. Once we had captured some footage of bony banging and slowed it down to sort out the chain of events, we became more and more convinced that Randy was right. From what we could make out, the dolphins swirled around at just the moment when the bang occurred. That plus observations of fish actually knocked out of the water fit Randy's hypothesis better than Ken's. We had to believe that the dolphins were hitting the fish with their tails.

So this wasn't the long-awaited proof of Ken's prey-stunning hypothesis, but it was still pretty amazing to watch. The dolphins worked hard to keep the fish schools contained, then took careful aim to deliver a good whack. It is not easy to smack a fast-moving fish. Once we knew what to look for, on many occasions in the shallows at Monkey Mia, we saw youngsters practicing, swiping their flukes at tiny fish, which they usually missed.

Bony banging was also an interesting social affair. Once, as we were following three males, Chop, Bottomhook, and Lamda, they suddenly took off, charging down to the south, leaping and porpoising in their haste to get to wherever they were going. We gunned the motor and followed them. When we arrived we found them snagging dejectedly on the outskirts of a bang-feeding group consisting of the Monkey Mia females, Nicky, Puck, Holeyfin, Holly, Crookedfin, Joysfriend, and Joy.

Why didn't Chop, Bottomhook, and Lamda join in the activity? They must have been hungry or they would not have gone charging down there in such a hurry. It looked as though they had arrived and then realized that they "weren't invited." As we turned our attention to the bony-banging females, I saw Crookedfin swipe at a fish. With a loud *bang*, the fish went flying and landed near another dolphin. This dolphin grabbed the fish and ate it. Crookedfin seemed completely unconcerned. Were they not only cooperating to keep the fish school contained, but also eating each other's fish? If so, that could explain why the males were not free to join in.

Later, I read an account of "pinwheel foraging" by killer whales off the coast of Norway. It sounded just like bony banging, and the researchers saw that the whale who hit the fish was not always the one who ate it.

Even as we found no proof that dolphins stun their prey with sound, evidence was amassing from studies of captive dolphins that they seemed to bombard their prey with trains of intense clicks, causing the fish to become disoriented and easy to catch: a sort of slow-acting version of the stun gun.

Whenever the dolphins hunted right in the shallows at Monkey Mia, shouts of excitement from the waterfront would draw us down to the shoreline. On one such occasion, as we trotted down the beach, Puck zoomed by at top speed just a few yards offshore, traversing the length of the camp in moments. A few chuffing breaths revealed her exertion. She then took off, heading east to pass right under the jetty and continuing off to the east, dorsal fin slicing through the water and bobbing with the motion of strong tail beats.

Andrew and I ran up the beach, carrying underwater recording equipment and videocameras, hoping to see what Puck was chasing and whether or not she caught it. She changed direction abruptly, then bucked to a stop facing into the shoreline in very shallow water. Then we noticed, between Puck and the water's edge, the shimmering form of a large yellowtail bream in only a few inches of water. The fish had sought safety from Puck by moving into water too shallow for her. But Puck was not to be dissuaded. She waited patiently just offshore of the fish, oriented toward it and barraging it with a relentless stream of intense echolocation clicks. The record level meter on my tape recorder kept swinging into the red, and I repeatedly adjusted it lower and lower. The fish seemed agitated, cruising along slowly, then making a sudden rush, turning this way and that, almost running right up onto the sandy beach. Finally the flustered fish took off in a panic. Puck was in pursuit, moving alongside the fish and then, as they came upon deeper water, behind it. As she passed by my hydrophone, Puck's echolocation clicks were so intense that I had to tear the headphones off my ears. The water resounded with an overwhelming, harsh, metallic clang overlaid by a high-pitched whine. I can only imagine what all that noise must have been like for the fish. Its lateral lines must have been twanging, its swim bladder fit to burst. No wonder it panicked.

Once more the fish escaped into shallow water. It seemed exhausted and disoriented and bumped right into my ankle. It paused, moved toward me, and again bumped into my ankle as if bewildered. It occurred to me that I could probably pick it up with my bare hands. I shifted my tape recorder bag over my shoulder and leaned over, gently closed my hands around the confused and exhausted creature, and lifted it out of the water. Trembling and slippery, it arched sideways, bending head and tail skyward as it stared into the brilliant sunlight. It seemed huge and heavy. Quite a substantial meal for a dolphin—or a human, for that matter.

Then something nudged against the back of my knee. It was Puck, mouth agape: she wanted to claim her prize. I laid the fish crosswise in her jaw, and she clamped down on it and turned away. Out in deeper water, she dove several times, swiping the fish against the bottom to break off the head. When Holeyfin approached her, Puck rolled onto her side with the fish protruding from her jaws as if showing it off. She lay there for a moment before diving once again. Next time we saw her, Puck was heading back over toward the jetty with Holeyfin. The fish was gone.

When we got back to the jetty, Holeyfin, Nicky, Puck, Snubnose, and Bibi were all gathered around, oriented toward the jetty. Their heads were scanning back and forth, typical of an echolocating dolphin. To confirm it, I put my hydrophone into the water and listened. The intense buzzing and clicking was deafening, like many electrical appliances shorting out. The dolphins were directing an onslaught of loud clicks directly at a school of bream that had taken refuge against the jetty pilings. The fish milled nervously, unwilling to venture farther than a foot or so from the pilings. The dolphins waited patiently, bombarding the fish with clicks, waiting for them to panic and bolt. Once away from the pilings and isolated from the rest of the school, individual bream would become easy targets.

Eventually one fish made a break, and Holeyfin took off after it. She had to make a detour around a boat mooring rope that the fish could fit under, but I knew she was still on top of it when a young boy on the beach started jumping up and down excitedly and pointing into the shallow water right in front of Holeyfin. She got her fish, as did each of the other dolphins that afternoon. Nicky had hers handed to her by one of the rangers, who, like me, was able to pick up the fish with his hands. The fish was either disoriented by the constant bombardment of dolphin clicks or perhaps just exhausted after being pursued up and down the beach.

I remembered the first time I swam with dolphins, off the coast of Hawaii. Some friends and I had been out sailing and stopped for a refreshing dip. Two large bottlenose dolphins came over toward me as I was swimming. Before I could even see them, I felt a funny tingling sensation and heard their buzzing echolocation. The sound seemed to enter my body and set all my organs abuzz. In fact, that was exactly what was happening. The dolphins' echolocation clicks are designed to transmit through water, which is what our bodies are composed of, for the most part. I was not frightened, but the sensation was certainly strange.

I can just imagine what it would be like to be the size of a yellowtail bream and have a behemoth dolphin following me around, vibrating the living daylights out of me with a deafening chain saw–like buzzer. It would be more than strange. After a while it could make one crazy and disoriented.

If dolphins can zap their prey with sound, what else might they be able to do with their buzzers? We had often seen dolphins buzzing at one another's genital areas, under the "armpits" (pectoral fin pits), or along their bellies. Researchers watching dolphins elsewhere and in captivity have also reported these behaviors. In fact, dolphins have receptors on various parts of their anatomy that appear to be especially responsive to the tactile effects of sound. Are they tickling each other or "just having a look"? Maybe both.

Having witnessed some of the phenomenal and innovative ways in which dolphins make use of their physical abilities, I would not be at all surprised if someone, someday, finds proof of the prey-stunning hypothesis and other equally amazing capabilities that we have yet to dream of. Just another reminder that, while we may share the same weather, we live in very different worlds and have vastly different sensory experiences.

DOLPHIN

SOCIETY

While the discovery of bony banging and of sponge carrying had contributed to my sense that dolphins are smart, creative, flexible, and innovative, we still had not gone far enough to satisfy my curiosity about what goes on in their minds. We had been watching and studying the dolphins for three years before the outlines of their complex society began to emerge clearly.

Andrew had returned with me to Santa Cruz after our field season in 1985. We arrived in California at the end of our financial rope, only to find my VW bus in need of major repairs and Ken's lab at Santa Cruz crowded and crazy. Ken had nurtured many young marine mammal biologists like ourselves, all of whom had come to rely on the resources of his lab. Some, like us, virtually lived in the lab, and there were times when the offices looked more like a living room/laundry room/kitchen than a research lab. A few people had taken to camping out on the floor, and their sleeping bags lay rolled up under a lab bench during the day. It was getting harder and harder to find the space and access to equipment that we needed during our short and intense periods of data analysis back in the States, and the management at the lab was cracking down on campers.

We had a lot of work to do: writing reports on our observations, getting our records in order and entered into computer databases so we could access them, and applying for financial support to continue our work. We had managed to stretch funds from the National Geographic Society over two years, complemented by our own savings. But we were broke now, and there seemed to be few prospects. We applied for further support from the National Geographic Society, which we thought to be our best bet. But so did Richard, who was ready to return to Monkey Mia after a hiatus spent doing the first years of coursework toward his Ph.D. Richard, with the support of his hard-hitting Ph.D. committee members on the faculty of the University of Michigan, got the grant, and we were out of luck. Only a few days before we planned to leave in the late spring of 1986, we learned that we had no support. We were depressed and anxious, having already put both feet "out the door," ready to go. It seemed we were on the verge of so many new discoveries and insights, yet we simply could not afford to return to Australia. In retrospect, I don't know how we pulled it off, but somehow we boarded the plane. Both Andrew and I convinced our families to lend us some money, just enough for the airfares, and we headed back to Australia with virtually empty pockets, knowing that at least our boat and tents lay waiting. We could eat fish and find some odd jobs to tide us over. At least we would get back to dolphin watching.

Some insights slowly dawn upon you, bit by bit, so there is no clear distinction between the time when you know something and the time before you knew it. Later it comes as a surprise to realize what you know. Other insights explode into your consciousness with the force of a partridge taking flight in the quiet of the north woods. I remember one of those latter moments well. We were on our way back to camp one evening after a long and exciting day of dolphin watching. It was one of those days when the bay seemed to be swimming with dolphins, and we had gone from group to group to group, tallying who was present and what they were doing.

By now we had succeeded in developing an almost complete cast of characters in Red Cliff Bay, the waters surrounding Monkey Mia. We rarely encountered dolphins we could not identify, and we had succeeded in sexing most of the dolphins we saw regularly.

The groups of dolphins we encountered seemed to be changing constantly as members joined and departed and groups merged and split. These comings and goings seemed random and chaotic, but the more we watched and the more each encounter was enriched by memories of previous encounters, sexes, identities, and habits of our subjects, the more we began to see patterns. At one level groups seemed to be fluid and constantly changing, but there were also some dolphins who tended to be together consistently.

On our way home on that particular evening, we kept encountering group after group of dolphins. Though we were exhausted and hungry and the light was failing, we couldn't pass them up without at least seeing who was present in each group. We came upon one group: Yan,

Surprise, Yogi, Square, Fatfin, Tweedledee, Crookedfin, Holeyfin, and Nicky. We had finally determined enough sexes that we knew every one of these dolphins to be female. After a short visit, we left them and promptly came upon another group: Chop, Bottomhook, Lamda, Trips, Bite, and Cetus, all males.

A mental review of most of the groups we had seen earlier in the day confirmed what suddenly seemed totally obvious. Not only did the dolphins have preferred partners, but those preferences fell out along the boundaries of sex. Males hung out with males, females with females. When we got back to shore we reviewed the groups we had seen that day and over the past several days. Almost all consisted of either all males or all females. My notebook entries from that day are full of exclamation marks and expletives underlined in red ink and starred.

The bay was not just a "soup" of dolphins, mixing and matching randomly with no particular rhyme or reason. The dolphins had definite, same-sex partner preferences. This was not just a vague impression or something we hoped might pan out in our statistical analysis. It was a slap-you-in-the-face obvious pattern. We had just needed to reach a "critical mass" with our sex determinations before it popped out at us, and today was the day.

We finally had a real pattern that begged explanation. What is the basis of dolphin relationships? Why do certain individuals hang out together, and why are they virtually always members of the same sex? The pattern seemed to hold even more for males than for females. The males formed little "gangs" of two, three, or four individuals who were basically always together. They might join up with other dolphins, but they always did so together. Red Cliff Bay was stomping grounds for several male gangs. We would get to know some of them well, including the threesome Chop, Bottomhook, and Lambda; another threesome Trips, Bite, and Cetus; and a foursome (or two pairs?), Realnotch, Hi,

Patches, and Hack. These gangs comprised fully adult, well-speckled (mature) males. There were also some gangs of mostly younger, less speckled males, including the threesome Lucky, Pointer, and Lodent and the pair Wave and Shave. These little gangs were so cohesive that we could pretty well be certain if we saw one member, the others were nearby.

Females didn't form such discrete gangs. A few, like Square and her daughters, plus Tweedledee and Fatfin, were pretty consistently together. But generally there were loosely defined networks of females who tended to be with each other more often than with others. Females hung out with other females but seemed to be less particular about which other females they were with.

Why would some male dolphins develop such close bonds with each other, spending nearly all of their time together? The question loomed large for us, because it is such an unusual pattern for a mammal. In almost all mammal species, males either avoid or barely tolerate each other. At best they might interact amicably on rare occasions. From a theoretical standpoint, why should they do otherwise? A single male can potentially fertilize many females. As long as the females can successfully rear the offspring he sires, his best strategy, in evolutionary terms (the one that will leave him with the most offspring), is to mate with as many females as possible and avoid having to "share" with other males.

For example, as I am writing this in rural Vermont, a fat little red squirrel is grooming himself on the deck by the bird feeder, which he has just emptied yet again. He lives alone, fiercely defending his bird-seed source and his den from other red squirrels, occasionally even fending off the much larger gray squirrels. He will, for a short time this spring, find it in his heart to be sociable, but only with female red squirrels. He'll fight with any other male who comes near his territory. A typical male mammal: solitary, territorial, intolerant of others of his species except for females, and even then only during the breeding season.

There are some mammals, unlike the red squirrel, that are social and live in groups, but for them the issue of males getting along together is moot since their groups consist of one male and several females. It is a rare few species that live in multimale/multifemale groups. Our closest relatives, the chimpanzees, are one example. They live in communities, often containing many males. Male chimpanzees spend an inordinate amount of time working out their relationships with other males, striving for status in a dominance rank. Dominance hierarchies are strictly enforced, and male chimps fight with each other, sometimes brutally. But in addition, males occasionally cooperate with each other, forming alliances to gang up against other males. For example, the beta male chimp may assist the alpha male to suppress any uprisings on the part of other males in the community.

Male chimpanzees also cooperate occasionally on a larger scale. Several males in a community will occasionally band together to go on "border patrols." They travel quietly along the borders of their territory, listening and looking for signs of chimpanzees from neighboring groups. When they come across a neighboring male, they may kill him outright. If it is a neighboring female, they may kill her offspring (fathered by one of the neighbor males) and injure her in the process, but she also may eventually be recruited into their community. In short, male chimps cooperate with each other to compete with other males, either from the same or from different communities.

Another mammal species that lives in groups with multiple adult males is the African lion. Female lions—usually sisters, daughters, cousins, and aunts—band together to form prides. They raise their cubs together, even sharing the responsibilities of nursing. Males also band together into groups of two to about six or so who work together to monopolize a pride. First they cooperate to oust the males currently residing with a pride, engaging in sometimes lethal battles. If they succeed, they might kill the pride's current batch of young cubs, sired by the old males. Then they settle in and wait for the females to come into estrus—the time when they are fertile and could become pregnant again—which will come sooner now that the females no longer have nursing cubs. Now the new males have their opportunity to mate and sire a new generation of cubs. Once again, males are cooperating in order to compete more effectively against other males.

Why do the female lions tolerate this, and why would they want to mate with the males who killed their cubs? The answer probably is that they simply don't have much choice. They may be able to exert some influence over the much larger males, but not a lot. It has been suggested that it is in the female's interest to mate with the victorious males. Given that this is the way of lion society, at least she will be raising cubs fathered by males who are successful, and her sons may inherit that talent.

Who gets to mate with the females if there are several males? Male lions who cooperate to monopolize a pride of females are usually brothers, cousins, or otherwise related. That means they have a lot in common, genetically speaking—about half of their genes in the case of brothers. The goal from an evolutionary perspective is to get those genes into the next generation. Whether they come from you or your brother doesn't matter so much.

In sum, wherever males do live together and form tight social bonds, the benefits of doing so come from cooperating in order to compete with other males. Chimps cooperate against males within their own communities and between communities; and lions cooperate to obtain and defend prides of females from other males. Knowing this and the basic theory behind male bonding, we knew something interesting had to be going on with the male dolphins in Shark Bay. But it would take considerable effort and a lot of dolphin watching to begin to sort it all out. Our first hints came in August of 1986.

Andrew and I were out surveying dolphin groups. We came across the male triplet Trips, Bite, and Cetus, with Yogi, a female, and the males Realnotch, Hi, and Patches. Booboo, Yogi's nearly independent son, was nowhere to be seen. They were incredibly wound up and excited as we approached, and the entire mob crowded together at the bow, whistling and wiggling around. One dolphin was belly up underneath another, petting and rubbing.

After a few minutes they grew bored with bowriding and went back to their business. Realnotch, Hi, and Patches were the first to leave and drifted back about one hundred feet behind the others. Trips, Bite, and Cetus were swimming in a rank abreast formation, lined up side by side and moving in perfect synchrony directly behind Yogi. Cetus suddenly rushed forward at Yogi, and there was some indecipherable splashing. Trips and Bite caught up, and they re-formed their rank behind Yogi.

Yogi was still ahead a few yards when we heard "knocks" (a sound dolphins make that is reminiscent of someone rapping their knuckles against a hollow log). The sound was coming from Cetus, whose head was partially out of the water. Yogi turned back and faced the males, then, with a kick of her tail flukes, she slid down underneath Cetus, rubbing the entire length of her body along his extended pectoral fin. When she got to her tail, she gracefully swept it upward and then swung back around to repeat the gesture. After three passes along Cetus, she settled into a snag alongside him.

Again there was a series of knocks. The water was only about eight feet deep, at the edge of a shallow weed bank, and we could see right to the bottom. Yogi dove, pointing straight down with her rostrum and poking into the weedy bottom, searching after a fish. But the three males immediately surged up at her in a fury, and she was forced to break off her hunt. Shortly thereafter there was another bout of intense splashing and chasing, mostly impossible for us to sort out, but when a dolphin's head broke through the water surface, we could hear a harsh, screeching, growling sound in air. Whatever they were up to, it looked and sounded aggressive. They certainly weren't just playing.

Eventually the dolphins all calmed down after a lot of circling and jawing and resumed traveling, with Yogi out front and the three males lined up, moving in perfect synchrony just behind her. Cetus then arched his head out of the water in a funny, awkward-looking position and swam in a tight circle alongside Yogi, bobbing his exposed head up and down in an odd, exaggerated motion. He seemed to be showing off to Yogi, strutting his stuff, like a peacock fanning its gaudy tail.

Every time Yogi made a move or the males suspected that she might, they charged at her, chasing, splashing, and making aggressive sounds. Then, just when things calmed down a bit, Yogi bolted. Taking off at full speed, she left a trail of flattened "fluke prints" on the water surface. Because the water was so shallow, the tip of Yogi's dorsal fin stuck out, streaking along, the tip bobbing up and down with each hard pump of her tail flukes. By the time she had gained a ten-yard lead, all three of the males were in pursuit.

We joined the fray, trying to stay close enough to see what was going on without getting in the way or running the boat aground. The males broke rank and spread out to get on either side of and behind Yogi. She zigzagged, trying to avoid them. The chase continued for about 150 yards before they caught up to her. A great wall of water flew into the air, accompanied by a dull thud as bodies collided. The details of what was going on were shrouded from view by a violent-looking splash and spray.

That observation of Trips, Bite, Cetus, and Yogi impressed us because of the intensity and aggressiveness of the males, but we were at a loss as to what it was all about. Making sense of dolphin behavior out in the bay was always a challenge. So much occurred underwater where we could not see, or the dolphins were moving around so fast that we couldn't stay close to them to see what was going on.

Therefore, when Snubnose and Bibi began some of the same shenanigans in the shallows at Monkey Mia, we were thrilled at the opportunity to learn more. A week or two earlier, they had come into Monkey Mia with a strange dolphin. They had been extremely excited, agitated, and distracted around this stranger, and it was unprecedented that a stranger should come right into the shallows among the tourists, having never before visited and probably never having been close to humans before.

A few days later they came in with the same stranger, whom we had dubbed Buster. Snubnose in particular was extremely tense and excited. He kept swimming right behind Buster, and now and then he would rush forward, positioning himself so that his face was poked into Buster's genital area, all the while emitting an intense buzzing-gurgling sound. He seemed to be inspecting the stranger's genital area.

Meanwhile Bibi was in among the people, taking fish handouts, but he was clearly distracted. After a minute or two in the shallows with his face out, mouth opened, he would snatch the fish and turn immediately to look out toward Snubnose and Buster as he gulped it down. Once he even broke away from the people and went out to join Snubnose and Buster. Taking up position alongside Snubnose, the two males rushed up and inspected Buster's genitals together. Buster turned over sideways and lolled at the surface—a female; I could just make out the mammary slits. She was fairly well speckled, too. Of course: sex. What else could possibly excite these males even more than food?

It was very out of character for Snubnose to miss out on an entire feeding session, but not once had he left Buster's side to go into the shallows where the tourists jostled their buckets of fish enticingly just a few feet away. He was clearly too preoccupied. Now, while he snagged alongside Buster, I could hear him making the *knock knock knock knock knock* sound. I had heard this sound many times before, but never in a context where I could also see what the dolphins were doing. Buster circled nearby. A moment later, more knocking and then more and louder and deeper.

Snubnose was working himself into a tizzy, tense and twitching, with the whites of his eyes showing. The knocks became louder and lower. Suddenly he swirled around toward Buster, lunging at her with his jaw open and making a horrible screeching, growling sound that left no doubt in our minds about how he felt. Pissed off. He and Buster began circling each other in very tense, tight little circles that gradually relaxed into larger circles, and then they snagged alongside each other, calming.

A little while later, when they all went chasing and splashing offshore, I slipped into wet suit, mask, and snorkel and swam out to them. Twice I watched as both Snubnose and Bibi approached Buster from either side, both with their penises erect, rubbing against her as they drifted out of my range of view.

They came back into the Monkey Mia shallows, and still Snubnose refused to leave Buster's side to approach the tourists. Even when he was not looking directly at her, he seemed to have his full attention focused on Buster, as if he were just waiting for her to try something. When a tourist waded into the waist-deep water to approach him, he tossed his head irritably and moved off a few feet to snag closer to Buster.

When the feeding was over with, and the tourists clambered back out of the water with empty buckets, Bibi came out immediately to join the other two. They all headed offshore, and we took our cue to run down the beach and get the dinghy. After throwing the anchor aboard, we pushed off from the shallows while trying to keep track of the dolphin's heading.

We caught up to them offshore and found Bibi and Snubnose side by side and just behind Buster's tail, following her every move with tense attention. Each time she made the slightest move, they rushed up at her, charging, chasing, splashing, and hitting, just like Trips, Bite, and Cetus with Yogi a few days earlier. I found myself annoyed with them, wishing they would leave Buster alone, but they continued to harass her for the rest of the day.

These observations were confusing. In retrospect I realize that this was in large part because of our preconceived ideas about dolphins. I still assumed they were always gentle and kind and playful. It never crossed my mind that they might have a darker side. The way the males had been so tense, aggressive, and downright nasty seemed out of character. Prior to these observations, we had called any roughhousing we saw play. But this was not playful, and there was no denying it. It was especially difficult to accept such behavior on the part of Snubnose. Watching him charge, chase, and hit Buster while making those harsh screams, screeches, and knocks was a shock. Was there really a mean glint suddenly perceptible in those soft brown eyes?

The three "tame" males — Snubnose, Bibi, and Sicklefin — beg for fish handouts from people in the shallow waters at Monkey Mia. (AUTHOR)

Dolphins and humans have interacted at Monkey Mia since at least the 1950s, and possibly much longer. (BARBARA SMUTS)

The author and Andrew Richards at "old camp." (Barbara Smuts)

A typical scene with dolphins and people in the shallows at Monkey Mia. (Author)

Andrew sitting on the bow of his boat, watching a bowriding dolphin. (PHOTOGRAPH COURTESY OF ANDREW RICHARDS)

Nicki-the-human offers a Long Tom fish to a dolphin. Long Toms skitter across the water surface on their tails in an attempt to escape capture. (AUTHOR)

The author recording dolphin vocalizations in the shallows at Monkey Mia.
(ANDREW RICHARDS)

Holeyfin and Holly, mother and daughter, playing keep-away with a piece of seagrass.
(AUTHOR)

Dolphins playing with seagrass.
(AUTHOR)

*Halfluke carrying
a large sponge.*
(AUTHOR)

*Nicky and Puck
"goose" Holly as
they all ride at the
bow of our boat.*
(AUTHOR)

Two alliances of males, Chop-Bottomhook-Lambda and Trips-Bite-Cetus, travel together in a "superalliance." Slight differences in interdolphin distance and surfacing synchrony indicate relationships. (ANDREW RICHARDS)

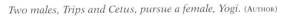

Two males, Trips and Cetus, pursue a female, Yogi. (AUTHOR)

Two males synchronously press their genitals against the sides of a female they are herding. (ANDREW RICHARDS)

A synchronous surfacing by two allied males. (ANDREW RICHARDS)

Two males, Realnotch and Hi, in rank formation behind the female, Puck. (AUTHOR)

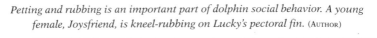

Petting and rubbing is an important part of dolphin social behavior. A young female, Joysfriend, is kneel-rubbing on Lucky's pectoral fin. (AUTHOR)

BOYS

Zzzzzzzzzzzz, zzzzzzzzzzzz, zzzzzzzzzzzz. Frozen in place, I was standing in knee-deep water in the Monkey Mia shallows, recording equipment hanging over my shoulder. Sicklefin had his open jaws poised around the calf of my leg and was emitting a crazy, screeching, cicadalike buzz that I had never before heard coming from any dolphin. In combination with the flash of his many fine, pointed teeth, I was reminded of a chain saw. This was obviously a test. If either one of us flinched, I would end up with a bloody, shredded leg. His teeth, though not particularly long, were sharp and plentiful, and I had developed a healthy respect for them. Bibi approached, attracted by the sound, and joined Sicklefin in his antics. Now both of them, tense and twitchy, eyes wide, ran their open jaws up and down my legs, Zzzzzzzz, zzzzzzz, zzzzzzz. They sounded like air raid sirens. I tried to remain calm, and after a minute or so they moved off, attention diverted by the approach of a tourist. Whew, saved by a bucket of fish.

It was the spring of 1987, and Richard and I had just returned to Monkey Mia together, having spent several months once again back in the United States. A friend had offered to pay my way back to Australia, where I helped him to restore and sell a house in Perth. But more than anything, I had longed to get back to Monkey Mia. Working as Richard's assistant was my ticket.

When we arrived, we found Sicklefin among the tourists. He had joined the ranks of the hand-fed Monkey Mia dolphins. In prior years he had come into the shallows occasionally and seemed fairly relaxed about people, but he never accepted fish handouts or allowed anyone to touch him. But now he was acting like an old hand: head out of the water, mouth open, he gulped down fish offerings held out to him by human hands in the blink of an eye.

From the ranger's reports and our own observations in the first few days, it became clear that Snubnose, Bibi, and Sicklefin together were coming in to Monkey Mia accompanied by an offshore dolphin almost every day. Some of the dolphins they brought with them were strangers to us. The intense following, chasing, genital inspections, and knocking sounds we had seen and heard once or twice during the previous field season were now a daily occurrence, and the dramas were taking place right here in the shallows at Monkey Mia. Now we might begin to make sense of it.

Over the following months we determined the sexes of many of the dolphins that came in with the males: all were female. Sometimes Snubnose, Bibi, and Sicklefin brought in the same female for a week or so, and at other times they came in with a different one from day to day. While in the Monkey Mia shallows, one of the males "stood guard," remaining farther offshore with the female and often forgoing fish handouts while the other males ate.

The knocking sound was incessant. Richard and I began to take note of the female's behavior in response to these vocalizations, and we soon realized that she would almost always turn toward and approach the knocking male. If she did not comply, or didn't do so fast enough, the knocking escalated into furious outbursts of screaming, accompanied by jerking head motions, charging, chasing, and hitting. It was as if he were saying "Stick by me or else. . . ."

The strutting display we had seen Cetus perform to Yogi turned out to be just one of a bizarre array of displays that the males performed. Many were performed by two males moving in perfect synchrony. They might start out side by side and behind the female ("flanking" her), then zoom up, one on either side, tilting their bellies toward her. Then they would both slap their chins down against the water surface, followed by an exaggerated tail-out dive, bringing their tails way out of the water and slapping the surface as they submerged. Underwater they would circle around to reapproach her, resuming the flanking position from which they had started. Or they might leap alongside her, each going in the opposite direction before circling back alongside her.

Another time I watched as both Snubnose and Bibi swam along on their sides next to each other, both wildly waving a pectoral fin in the air. They looked just like circus dolphins performing a show trick. The sheer variety of these stunts was mind-boggling, especially so because they were performed in such a perfectly coordinated way. How did they do it? There was no obvious leader and follower. Did they somehow discuss what they were going to do beforehand? It was difficult to tell much of the time whether these displays were performed for the sake of impressing the female or for the sake of the participating males themselves. Sometimes they went through elaborate displays even when there was no female around, leading us to suspect they might serve to demonstrate and enhance male-male solidarity.

When we followed them as they traveled offshore, away from the Monkey Mia shallows, Snubnose, Bibi, and Sicklefin traveled in perfect synchrony, surfacing and diving together in a rank abreast formation directly behind the female. Even in the shallows, the males moved with precise coordination, as if they were linked by some invisible electronic connection. Snubnose, Bibi, and Sicklefin were united in their purpose, and their purpose, apparently, was to impose themselves upon these females.

Richard and I settled into a routine. In the morning we watched Snubnose, Bibi, and Sicklefin with whichever female they brought with them, and in the afternoon, weather permitting, we headed offshore to check up on the offshore dolphins. Most days we encountered the Red Cliff Bay male gangs (Chop, Bottomhook, and Lamda; Trips, Bite, and Cetus; Realnotch, Patches, Hack, and Hi; and the others). Usually one or more of these gangs also had a female with them. Out in the deeper water, we couldn't see what was going on below the surface as well, but what we did see fit the same patterns we were observing with Snubnose, Bibi, and Sicklefin: males traveling in a synchronous rank abreast behind the female, the knocking sound, occasional charges, and hitting and splashing. If it hadn't been for Snubnose, Bibi, and Sicklefin carrying on in the shallows at Monkey Mia, it might have taken many more years for us to begin to make sense of the behavior of the males.

Richard was the first to really understand the theoretical implications of what we were seeing. He had been immersed in evolutionary theory at the University of Michigan and so was primed to think about animal behavior in terms of the costs and benefits (in terms of reproductive success) of conflict and cooperation. He recognized how unusual it is in the animal kingdom to find males cooperating with one another, and his excitement and enthusiasm about it were contagious.

Later during that summer we had a visit from three scientists, all of them among my all-time heroes in the study of animal behavior. Irv DeVore, chair of the Anthropology Department at Harvard, had pioneered the study of nonhuman primates in the early 1960s, as well as ethnological studies of the !Kung bushmen of the Kalahari and Pygmies of Central Africa. Richard Wrangham, a professor at the University of Michigan, had studied the ecological bases of behavior in chimpanzees at Jane Goodall's Gombe Stream Reserve and had just received a MacArthur Award. Barb Smuts, also a professor at the University of Michigan, had worked at Gombe and had written a fascinating and important book entitled *Sex and Friendship in Baboons*. These three visitors injected our dolphin watching with a new energy, a sense of validation, and leadership in developing solid scientific methods for observing the dolphins.

Richard Wrangham and Barb Smuts were on Richard Connor's Ph.D. committee. Excited by what they saw of the behavior of the male dolphins, they suggested that he focus on male behavior for his thesis. They agreed to help Andrew and me to get into graduate school and develop our own thesis research, but they felt we needed to slice up the research pie to avoid overlap and potential interference so that ultimately we could all complete good, focused doctoral research. I was fascinated by the complexities of the behavior of males, and we were clearly poised on the brink of some important discoveries. So it was with some reluctance that I accepted this arrangement. I eventually decided to focus my thesis research on the dolphins' acoustic communication, and Andrew focused on female social relationships. The project took on a new dimension now with independent projects and more clearly defined research boundaries and goals. But I still spent much of my time watching males, usually by collaborating with Richard.

During the next couple of years, virtually every day revealed new and yet more amazing complexities to the dynamic relationships among males. We eventually referred to the stable male gangs as "alliances" because of the cooperative nature of their behavior. The males cooperated to "herd" females. Herding typically began with the males choosing a female, often separating her from her companions and even chasing her aggressively. They performed displays around her, swam in a tight rank formation just behind her, and frequently made the knocking vocalization. Sometimes they inspected her genital area or mounted her. Herding was often an aggressive affair. The males would jerk their heads and scream at the female they were herding, and frequently that escalated into chasing and even hitting her. The female usually ended up staying with the males for anywhere from a few hours to a few weeks.

Most male alliances were triplets, with some pairs. Yet when they were herding a female, only two of the three members of a triplet seemed to be really involved. The third rarely made the knocking sound and made no special effort to remain close to the female. He was unlikely to participate in the synchronized displays or take up rank formation behind the herded female. Instead he hung around on the outskirts of the activity, usually foraging: the odd man out.

If the female bolted, trying to escape from the males, the odd man out would join the chase, but otherwise he seemed to be just standing by. Each time a triplet herded a different female, the odd man out might change, although some males were more often the odd member of their triplet.

It was easy to tell who was odd. Males typically traveled together in a rank abreast formation, surfacing to breathe in perfect unison, except for the odd man out, who was just out of step and often slightly farther apart than the other two. Just by watching how the dolphins moved with respect to one another, we could discern something of their relationships.

We also noticed that each of the male alliances we knew in Red Cliff Bay cultivated a special relationship with another alliance. Snubnose, Bibi, and Sicklefin spent a considerable amount of time with Wave and Shave. Trips, Bite, and Cetus spent a lot of time with Chop, Bottomhook, and Lamda and, more recently, with Realnotch and Hi. These second-order alliances traveled together or nearby, and they often petted and formed synchronous surfacing pairs involving males from the two different alliances. Occasionally males from second-order alliances even paired up to herd a female together.

It was remarkable enough that males cooperated in pairs and triplets, but why should they then also form bonds with other pairs and triplets? One particularly memorable day of dolphin watching revealed insights into the politics of male alliances that we had never imagined.

It was August 19, 1988, a day that none of us will ever forget. It looked from the onset to be a good boat day: calm and clear. But before heading out on the boats, we wanted to see what the Monkey Mia dolphins were up to. Walking down to the beach that morning, Richard and Andrew and I could see from a distance that something was up. Snubnose, Bibi, and Sicklefin were in, and whatever was going on, it was exciting, with lots of splashing and fast action. I could hear dolphins squealing even way up on the beach.

When we got down to the water's edge, Dave Charles, the ranger on duty, filled us in. "Holeyfin is back, and the guys are pretty keen on her." Indeed, the males were paying no attention whatsoever to the twenty or so tourists who were wading in and out of the water with fish in hand. Instead they were surfacing in a tight, tense, synchronous rank behind

Holeyfin. She hadn't visited the Monkey Mia shallows for the past several days, and we had seen her several times offshore with the male alliance of Chop, Bottomhook, and Lamda.

Holly was four years old now, old enough to be pretty much independent of her mother, so Holeyfin was probably ready and able to get pregnant again. By examining our records, we determined that herded females were usually those who were not pregnant and didn't currently have young infants: precisely the females in the population who *could* become pregnant. Holeyfin was ready, and she had been attracting a lot of attention from the males in the bay lately.

Female dolphins are seasonally polyestrous, meaning that during a broadly defined breeding season—the austral summer in Shark Bay—females go through several hormonal cycles in which they ovulate and become fertile. They continue to go through these estrous cycles until they become pregnant. Because they have a one-year gestation period, the birthing season coincides with the mating season. We have no idea how males know, but somehow they are able to discern when a female is cycling. We had seen males "inspecting" the genitals of females they herded, and we suspected that they might be capable of "looking inside" with their echolocation and somehow assessing the state of her reproductive organs.

In many species of mammals, females exude distinctive odors or pheromones that signal to males when they are able to become pregnant. Among primates, many species sport huge red-and-purple "sexual swellings" of the labia and surrounding tissues that come and go with the female's cycle, peaking during the day or so when she is ovulating and most likely to become pregnant. Our closest relatives, the chimpanzees, have particularly grotesque and exaggerated sexual swellings. Female chimps at the peak of their cycle have trouble sitting down, their bottoms are so swollen. Though chimpanzee sexual swellings appear hideous to our eyes, male chimpanzees find them highly attractive, and they have presumably evolved as a means for females to signal that they are ready to become pregnant.

We differ markedly from our chimpanzee cousins when it comes to signaling estrus. We have no external signals, no apparent odor or visual signal. In fact, neither men nor women know when a woman is ovulating. But male dolphins, whatever the means, seem to know when to court a female. No sooner was Holeyfin back from her week-long adventure offshore with Chop, Bottomhook, and Lamda than Snubnose, Bibi, and Sicklefin took an interest in her. I felt a bit sorry for the old girl. Perhaps she was enjoying all this attention, but then again, after several days offshore away from her usual routine, she might just be tired and hungry. Snubnose, Bibi, and Sicklefin were not exactly easygoing company.

Followed closely by the males, Holeyfin joined Puck and took up a position right alongside her, so close that their pectoral fins touched. As Sicklefin snagged at the surface, Holeyfin and Puck both took turns diving and tilting alongside him so that his pectoral fin rubbed down the length of their sides. They each took two passes, while Sicklefin hung motionless, seemingly aloof or perhaps just blissed out. The females then snagged alongside him. We had seen females rub like this along a male's pectoral fin many times before, and I was always unsure whether she was being genuinely affectionate or just appeasing him, trying to impress him with her undying devotion in an attempt to avoid escalated harassment and aggression.

Snubnose and Sicklefin followed every move Holeyfin made, staying close to and usually just behind her. Twice they rushed up toward Holeyfin from behind and tilted sideways as they dove underneath her, both angling their heads toward her genital area and buzzing intently, inspecting her genital area.

Snubnose did the strut display, swimming tight circles around Holeyfin, arching his back so that his head jutted upward out of the water and bobbing his head up and down rhythmically so that his chin slapped down against the water with each stroke. While doing this, he emitted a steady stream of air through his blowhole: a gurgly, buzzing sound. To me he looked pretty ridiculous, but I suppose he figured he looked pretty cool. Maybe Holeyfin was impressed.

Richard and I were busily absorbed in recording our observations. I had a hydrophone in the water, recording the dolphins' vocalizations, and Richard was babbling as fast as he could into his tape recorder, describing the dolphins' behavior. During a momentary lull in the activity, we looked up from our work long enough to realize that two other dolphins were present. It was Trips and Bite, two of the males from the Trips, Bite, and Cetus triplet alliance. This was a rare event. Only once before had we seen Trips and Bite so close to shore at Monkey Mia. They were offshore dolphins, yet here they were, just twenty feet from shore. They floated quietly, oriented directly toward Snubnose, Bibi, Sicklefin, and Holeyfin, who seemed completely oblivious of their presence.

I waded out toward Trips and Bite in hopes of getting an opportunity to record some sounds from them. Even when I was right next to them, they showed no sign of acknowledgment. They just hung there, bobbing gently on the rippled surface. The sun glinted off their exposed foreheads, and I was close enough to see that they both had their eyes squinted partly shut. It was a thrill to be in the water so close to these wild males. I had often hung over the bow of our boat as they rode just a few inches from my face, but somehow being in their element with them made me more aware of their size and power. They were completely silent and almost gave the impression of being at rest, except that they kept turning slightly, readjusting to remain oriented toward Snubnose, Bibi, Sicklefin, and Holeyfin. They were obviously "casing the joint."

When, after about half an hour, Trips and Bite headed offshore, not in any great rush but traveling directly, surfacing side by side in synchrony, Richard backed out of the water. "Something's up. I'm gonna follow them out." Andrew and I agreed to stay on shore and keep an eye on Snubnose, Bibi, Sicklefin, and Holeyfin. Richard quickly loaded up his boat and headed out after Trips and Bite, with his assistant, Eric, at the wheel. I watched the boat grow smaller as they headed out to the northern horizon.

Snubnose, Bibi, and Sicklefin were following Holeyfin around, displaying and carrying on like a gang of punks harassing the neighborhood nerd, when Richard's voice came in over the radio. "Look out, here we come." Looking out toward the north, we saw Richard's boat not one hundred yards from shore, and just in front of the boat were at least five or six dolphins, in a wide rank, surging toward shore like a tidal wave. One or two of the dolphins leapt out of the water as they came rushing in to the Monkey Mia shallows. Then all hell broke loose.

As they swept in to the shallows, I heard horrible nasty growling and grunting sounds, like a group of lions attacking a family of warthogs. Snubnose, Bibi, Sicklefin, and Holeyfin took off at top speed to the west, and the others followed. A huge chase ensued, but they were all moving so fast that it was difficult to tell who was chasing whom. I heard the dull *thwack*ing sounds of dolphins hitting each other, but soon they were out of range of my hydrophone and away down the beach.

Richard and Eric followed along, furiously trying to keep track of who was where. Finally Richard reported back over the radio. "This is like science-fucking-fiction!" he exclaimed. "Trips and Bite went out north and joined up with Cetus, and then the three of them joined up with Realnotch and Hi, who are herding Munch. Then all of them together came steaming straight back in here and stole Holeyfin from Snubnose, Bibi, and Sicklefin. She's with them now, and Snubnose, Bibi, and Sicklefin are hanging back behind."

It was beginning to make sense now. Trips and Bite had come into Monkey Mia and assessed what was going on with Holeyfin and Snubnose, Bibi, and Sicklefin. Perhaps they had realized that they would need help in order to steal Holeyfin, so they had gone out to find their third member, Cetus. They had been spending a considerable amount of time with Realnotch and Hi lately, and they had apparently recruited their help as well. With all that support, they had launched their attack. The chase had gone on for quite a while, occasionally breaking down into bouts of fighting, hitting, ramming, and then more chasing. Although it had been hard to keep track of all that had transpired, one thing was clear: At the end of it, Holeyfin was now with Trips, Bite, and Cetus.

Realnotch and Hi had joined up with Trips, Bite, and Cetus to help them abscond with Holeyfin, even though they were already preoccupied with herding Munch (who remained with them throughout all of this). Once Holeyfin was secured with Trips, Bite, and Cetus, the two alliances had parted company, but not too far away, apparently remaining within calling distance. For the first time, we were beginning to understand why alliances formed special bonds with other alliances (what we referred to as second-order alliances). Two alliances working together could overpower a third alliance and steal a female from them or put up a solid defense against attempts from other alliances to interfere with their herding.

Cooperation between alliances added a significant level of complexity to the political world of male dolphins. Males had to sort out relationships not only within their alliances, but also with the members of the other alliance with whom they cooperated in a second-order relationship. This multilevel cooperation in dolphins was perhaps the most important discovery our research team would make. Cooperation among male mammals even on an occasional basis is rare enough, and alliance formation, where males form long-term cooperative bonds, is even rarer. But nothing as complex as long-term alliances among pairs and triplets forming long-term alliances with other pairs and triplets has been found in any other mammal besides dolphins and humans.

Among humans living in hunter-gatherer societies, for example, two brothers living in a village may share resources and help each other out. In turn, they may also develop cooperative relationships with a few other small groups within their village. All of these men may cooperate at times to defend their shared village. Several villages may sometimes band together as a united front against other similar bands of villages or neighboring tribes, and so on. Carried to an extreme in modern society, layer upon layer of cooperating bodies has resulted in massive institutions such as nation-states.

A group of males we referred to as the "wow crowd" later showed us that we had not yet exhausted the possibilities for complexity in male dolphin relationships. Numbering about fourteen, the Wow Crowd didn't fit the alliance formation pattern we had become accustomed to. Whenever we encountered them, the Wow Crowd males were all together. Pairings within the huge group were obvious, but they changed from one encounter to another. There were no stable pairs or triplets and no clear second-order alliances. Richard decided to try to sort out what was going on. After several field seasons devoted to watching them, he decided that the Wow Crowd constituted a "superalliance." They cooperated with one another, as did alliances and second-order alliances, but pairings within the superalliance were unstable.

Amazed as we were by the discovery of alliances and then by the discovery of alliances cooperating with other alliances, we had not yet (and perhaps still have not) learned all there is to know about the diversity of ways in which Shark Bay males cooperate.

After losing Holeyfin, Snubnose, Bibi, and Sicklefin hung back, agitated and apparently defeated. Andrew and I jumped into our boat now as well. We would follow Snubnose, Bibi, and Sicklefin, while Richard stayed with Trips, Bite, Cetus, and Holeyfin. When we caught up to them, Bibi and Sicklefin were petting each other furiously. It was as though they needed to reassure each other after such a defeat. The analogy with human behavior was downright comical, and we couldn't resist anthropomorphizing, putting words into their mouths: "Yeah, we put up a good fight, we're still the toughest . . . we'll get Holeyfin back from those lousy thugs, yaa: you're okay, Sicklefin, yaa, you're okay, Bibi. yaa . . . yaa . . ." If they were human, they would have been clapping each other's shoulders. The three were following along behind Trips, Bite, Cetus, and Holeyfin but keeping their distance. They came to the boat and rode at the bow, and I could see how excited and agitated they were by the quick tension of their movements.

Snubnose started to drift back behind the other two, and Bibi broke away from Sicklefin to go back to him. It seemed as though Sicklefin and Bibi were rallying the forces to pursue Trips, Bite, and Cetus, but Snubnose, in spite of his initial enthusiasm about Holeyfin, had little interest. Bibi seemed to be the most gung ho funny, given that he had been the least involved in courting Holeyfin back at Monkey Mia. With the little added encouragement from Bibi, however, Snubnose soon caught up, and the three of them began to surface synchronously, closing the distance between themselves and Trips, Bite, Cetus, and Holeyfin.

I could imagine them saying, "Ya, c'mon, let's get 'em . . . c'mon, Snubnose, don't be a chicken." But when they got within about forty yards of the others, they seemed to lose their nerve. They slowed, fell out of synchrony, and milled around nervously, and once again Snubnose drifted back behind them. Sicklefin and Bibi took to petting each other furiously again. "Looks like they've lost their nerve. Especially Snubnose. They don't really want to mess with Trips, Bite, and Cetus anymore," I reported to Richard over the radio.

Twice more, Snubnose, Bibi, and Sicklefin seemed to rally up the gumption to go after Trips, Bite, and Cetus, then lost their nerve when they got close. Finally Snubnose did a tail-out dive, heading away from the other two males. He had apparently made a decision. He preferred foraging to fighting. All three turned back toward Monkey Mia. Their spirits seemed to lift as they approached the shallows. A horde of tourists eagerly awaited the arrival of dolphins with buckets of fish. The males sped up as they approached and then parked themselves in the shallows, bellies to the sand, bracing with their pectoral fins, heads uplifted, mouths agape, as they were enthusiastically stuffed by their human admirers.

I couldn't help being struck by the disjunction between what people assumed about these dolphins and the reality of their lives. One lady kept referring to Bibi as "she," as in "Isn't she sweet!" To me, Bibi was probably the dolphin to whom that adjective was least applicable, and certainly the battle he had just fought was hardly "sweet." "She has such gentle eyes." To me they looked wild and mischievous, slightly bloodshot . . . unpredictable. This woman also got it into her mind that Sicklefin was Bibi's baby, though Sicklefin was at least as large as, if not larger than, Bibi. Perhaps she had heard that there was a mother and baby (Holeyfin and Holly) and then assumed that she *must* be seeing a mother and baby.

In any case, she kept telling Sicklefin he was a "gorgeous little imp." When Sicklefin snapped at her dangling hand, she laughed and told him he was being "awfully cheeky." Same planet, very different realities, I thought. Good thing these tough-guy dolphins don't understand what you're saying, lady!

After the feeding, Bibi and Sicklefin still seemed to be hanging tight together, with Snubnose on the outs. All three males moved off into deeper water, Bibi with his pectoral fin rested against Sicklefin's side. Snubnose joined, placing his pectoral fin against Bibi's side, but

Bibi turned away from Snubnose and started rubbing his body along Sicklefin's pectoral fin, breaking contact with Snubnose. Snubnose drifted apart. All three snagged at the surface.

A moment later Snubnose broke rank, lunged forward, circled around, approached Bibi and Sicklefin head-on, and pushed his way in between them, almost as though he were trying to break up the cozy camaraderie that excluded him. He rubbed his body along Bibi's pectoral fin. Bibi remained snagging, aloof, pressed close to Sicklefin.

Eventually all three males settled into a long bout of snagging, oriented out to the north. Perhaps they were tired after the morning's excitement and a belly full of fish. Perhaps they were listening to the sounds of dolphins in the distance. They began to travel out that way. Bibi petted with Snubnose briefly, then quickly rejoined Sicklefin, surfacing in perfect synchrony with him, while Snubnose remained a bit apart and out of step. I felt a bit sorry for Snubnose.

Bibi and Sicklefin, riding just in front of the bow of the boat, tilted on their sides, bellies toward each other, both twitching their heads in funny, jerky little motions. They did this several times, then approached each other rapidly, tilting, then butting their shoulders together, rolling so that the point of contact moved down from shoulders to tail stock. As they pulled apart, both males flicked their tails up and down in a fast, tense, exaggerated little flailing motion. They repeated this strange ritual several more times, slamming shoulders, rolling their sides together, then flailing their tails as they separated again. They reminded me of football players head butting after a good play. Yet another bizarre, ritualized display, this time clearly directed to one another rather than to any female. Perhaps some sort of "war dance" intended to develop solidarity and rile up the forces.

Sicklefin and Bibi, more and more excited by their little "dance," sped up, traveling faster and faster until they were leaping out to the north, Snubnose trailing behind. We followed along, gradually realizing that there was a fourth dolphin now, and they were chasing it. All three males porpoised in synchrony behind the new dolphin. The chase didn't last long, and when we caught up, we discovered the new focus of their attentions: Poindexter, a female.

As we caught up to them, Sicklefin and Bibi were displaying together on either side of her, arching their backs and bobbing their chins up and down, then slapping their tails and diving underneath her. A moment later Poindexter was belly up underneath Bibi, apparently rubbing her belly vigorously against his. Were they mating? We couldn't be sure.

When Poindexter started rubbing and petting with Sicklefin, Bibi briefly approached and petted with Snubnose. Poindexter then rubbed underneath both Bibi and Sicklefin, just as Holeyfin and Puck had done earlier with Sicklefin. They snagged side by side, touching each other's pectoral fins, seeming to pretty much ignore Poindexter.

As was so often the case, it was hard to tell how much all of this had to do with Poindexter and how much had to do with the relationships among the males. The males behaved as though they were compelled to prove themselves somehow after having been defeated by Trips, Bite, and Cetus. Herding Poindexter was their solution. Since they had lost Holeyfin a couple of hours earlier, they had been almost continuously displaying to each other and petting. Andrew and I, once again unable to resist the opportunity to anthropomorphize a bit, chuckled over the parallels between the behavior of these dolphins and that of a gang of nineteen-year-old human males, vying for status, showing off, challenging each other's manhood, reinforcing egos, driven to greater and greater displays of silliness.

Shortly, the males and Poindexter began traveling back south toward Monkey Mia. We were at least a mile and a half from Monkey Mia, and the tide was low. They would have to skirt around the edge of the shallow weed bank to stay in deep enough water. As they traveled, all of the dolphins dove down to the bottom and rubbed their bodies against the "trees" of seagrass that stood up from the bottom. Sicklefin sensuously rubbed his side along a seaweed tree, then hooked it with his tail, dislodging it from the bottom and trailing it along for a while before letting it go with a twitch to drift off behind him.

Sicklefin and Bibi continued to pet each other occasionally, sometimes even as Poindexter was rubbing both of them. At one point she went upside down underneath Sicklefin and rolled her entire body to and fro so dramatically that the entire width of her belly and sides swept back and forth against his pectoral fin. He simply held his pectoral fin in position while she did all the work. Again I had to wonder, Was this an indication of her genuine enthusiasm about Sicklefin or was she just trying to appease him somehow? Meanwhile Snubnose followed along behind, occasionally coming to ride the bow of our boat— cheap substitute for dolphin companionship.

The males brought Poindexter back to Monkey Mia. On the whole, and certainly compared with the way they had behaved with Holeyfin earlier in the day, they were disinterested in Poindexter. No displays, no following her every move, no surfacing in synchrony right behind her, no genital inspections. Were they just worn out from their exciting morning or was Poindexter just less desirable to them, and if so, why? Later in the day, Poindexter took off, and the males didn't even bother to pursue her. Over the next few days and weeks, Snubnose, Bibi, and Sicklefin often herded Poindexter, but we had the distinct impression that they were halfhearted. Poindexter seemed to be the fallback female.

What seemed most astonishing about the chain of events on August 19 was the way that Trips and Bite came into Monkey Mia in the morning and assessed the situation, then, in a way that suggested considerable forethought and premeditation, went out to recruit their third partner, Cetus, and their second-order alliance partners, Realnotch and Hi. What had gone through their minds as they watched Snubnose, Bibi, and Sicklefin courting Holeyfin that morning? Had they been scheming an attack? What and how had they communicated their plan to Cetus and to Realnotch and Hi?

Every day we observed fascinating new twists that shed further light on the emerging picture of male alliance politics and raised more and more questions. Why did so many alliances consist of threesomes? In a simple and ideal world, a male dolphin would do his own thing and avoid having to share mating opportunities at all. Sharing those opportunities with another male would be costly indeed, but perhaps he couldn't achieve any mating opportunities at all without some help. But splitting them three ways? This was terribly costly and puzzling given that the odd-man-out phenomenon suggested that two was a better number anyway.

We reasoned that considering the slow reproductive rate of female dolphins (one baby every four years, at best), only very few females are cycling at any one time. So fertile females are at a premium. Because of this, males must compete intensely for rare opportunities to mate and sire offspring. Such competition for mating opportunities is well documented in many animal species and is responsible for the fact that males in so many species fight viciously and evolve features like huge canine teeth and large size that enhance their fighting ability.

Perhaps the most extreme example is the elephant seal. Male elephant seals are huge, weighing several times as much as females and sporting fearsome canines. Males are so much larger than females that females are sometimes injured during mating. During the mating season, females haul out onto select beaches in large numbers. Males compete with one another viciously. Only one male will win, and he will win the entire harem of females, siring tens of pups during that season. The losing males will sire none (or close to it). With odds like that, competition is brutal.

Among dolphins, a lone male would be hard-pressed to mate with an unwilling female. All she has to do is be passive, and a male pushing up against her will succeed only in causing her to drift away from him. The only way he could force himself upon her would be to pin her against something. Among right whales, mating involves several males and a female. The female, surrounded by eager males, finds herself in the position where, as she rolls away to evade one suitor, she rolls toward another. Though the males aren't actually cooperating, the fact is that the female is pinned among them, and none would succeed in mating without the presence of the others.

Male dolphins may have started out in a similar situation. More than one male was needed in order for any one male to mate successfully with a less-than-willing female. This could have set the stage for cooperation. Two males working together had the tremendous advantage of being able to repel other males. Even though each member of the pair halved the chances that he would sire offspring with a female, the odds may still have been better than in a free-for-all, with many other males involved.

Of course, once a cooperating pair of males succeeds in outcompeting single males, others will be better off forming pairs as well. Pairs become the norm, and the advantage of pair formation is lessened. But what if one pair now adds a third cooperating member? The triplet now has the advantage of numbers, and their opponents will now be forced to take on third partners, so triplet alliances become the norm. Now, if one alliance forms a second-order partnership with another alliance, they have in effect doubled their numbers and forced their opponents to do the same. Evolutionary biologists refer to such escalations as an "arms race." I suppose we should thank the gods that male dolphins do not have the means to construct nuclear weapons.

Virtually every male dolphin in our study population belonged to an alliance. The only exceptions we saw involved males whose alliance partners died. Belonging to an alliance is the *raison d'être* for male dolphins in Shark Bay, and they get started early in life. Cookie, Smoky, and Jesse, for example, are three young males, all born within a few years of one another. They have spent quite a lot of time together since infancy, perhaps simply because their mothers share similar ranges, bringing them into regular contact. They roughhouse and cavort around as young mammals tend to do, while their mothers forage nearby. Since dolphins are so long-lived, it will be many years before we will know for sure whether they will grow up to be alliance partners, but I bet they will.

Although many of the alliance partnerships that existed when we started watching the Shark Bay dolphins in 1982 (we photographed them together that year) are still in place all these years later, not all have remained so stable over the long term. Several of the alliances that taught us about male behavior during the late 1980s have disappeared. When we first began working in Shark Bay, three alliances seemed to dominate Red Cliff Bay: Trips, Bite, and Cetus; Chop, Bottomhook, and Lamda; and Realnotch, Hi, Hack, and Patches. Over the years, Hack and Patches disappeared, leaving Realnotch and Hi to carry on. Trips, Bite, and Cetus all disappeared; and Chop and Lamda disappeared, leaving Bottomhook (who later joined Realnotch and Hi).

We have no idea what happened to all those males who disappeared. Perhaps they all died of natural causes, and then again, maybe not. A few times we have wondered if males might sometimes fight to the death or perhaps force other males to seek residence somewhere else.

The disappearance of Patches, a former member of the Realnotch, Hi, Hack, and Patches foursome (probably a second-order alliance of the two pairs: Realnotch with Hi and Hack with Patches), was particularly suspicious. One day we watched as Patches's partners, along with the alliance Trips, Bite, and Cetus, ganged up on him ferociously. When we first encountered the group, Hi was sidled up alongside of Patches with an erect penis, apparently a dominance display not unlike those performed by dogs. There was a lot of jostling and commotion, when suddenly a clear alignment formed out of the chaos. All six males lined up face-on against Patches. After a tense and foreboding moment, they charged at him, and we caught a glimpse of one dolphin biting at him and another striking with its tail flukes. A chase ensued, and when we caught up, the six had lined up against Patches once again and attacked him a second time. Afterward Patches lay on the water's surface, neck arched, holding his head upward in a stiff and awkward-looking position, the whites of his eyes showing, emitting a shrill, squealing sound. He was scared and in pain. The following day we came across Realnotch, Hack, Hi, and Patches together once again, and all had fresh new wounds: a gash over Realnotch's eye, a new piece out of Patches's dorsal fin.

This event may well have been one of a series that eventually led to the dissolution of the second-order partnership between Realnotch-Hi and Hack-Patches and perhaps also the dissolution of the Hack-Patches alliance. A month or so after the attack on Patches, Hack disappeared, and not too long thereafter, during the hiatus between field seasons, Patches also disappeared for good. We never saw either again and were left to wonder about their fates.

It is impossible to know why Patches was ostracized so severely by his alliance partners, assisted by Chop, Bottomhook, and Lamda, but his attempts to forge bonds with Trips, Bite, and Cetus might have had something to do with it. Just beforehand, we had encountered Patches traveling with Trips, Bite, and Cetus one afternoon, far from his alliance partners. This seemed mildly unusual at the time. After the attack, we again found him without his alliance partners, in the company of Trips, Bite, and Cetus. Whether cause or effect, Patches was clearly currying favor with them around that time.

For many, the story of male dolphins cooperating and competing among themselves to herd females, aggressively forcing the females to stay with them and mate, is difficult to accept. We first published a couple of scientific papers on male cooperation and herding in scientific journals. Then, perhaps because it was so different from expectations or because such "politically incorrect" behavior on the part of intelligent males held such a horrid fascination, the popular press picked up on the story. Richard, as lead researcher on this topic, was inundated by requests for articles, filming dates, and interviews. *The New York Times* ran a long article entitled "Dolphin Courtship: Brutal, Cunning and Complex," and a PBS *Nova* documentary was produced called *The Private Lives of Dolphins.*

People found it difficult to swallow the notion that dolphins, supposedly the kind, smiling, gentle denizens of peaceful seas, could possibly be so rude and calculating. Elizabeth Gawain, the gentle soul who first told us about Monkey Mia, was one person who reacted with horror to our first reports on male alliances herding females. When she visited us in Australia in 1988, Snubnose, Bibi, and Sicklefin were in full swing, herding females in the shallows at Monkey Mia on a daily basis. She watched and listened to our interpretations of their behavior and questioned everything thoroughly, always seeking alternative and "nicer" explanations for the behavior of the males. But after watching Bibi and Sicklefin viciously attack a female they were herding one afternoon, she broke down in tears and told me that, much as she hated to admit it, she felt that we were right.

Elizabeth had the bigness of heart, mind, and soul to see the dolphins for herself and to do so open-mindedly. Ultimately she was able to accept what was eminently obvious. Others were far more resistant. One man wrote a master's thesis, "deconstructing" our research results and claiming that we were simply projecting our own personalities onto the dolphins, implying that other folks with better dispositions might have discovered the dolphins to be the kind and gentle, perpetually sweet creatures they are *supposed* to be.

More recently, dolphins (not in Shark Bay) have been observed to kill harbor porpoises, apparently for no reason, and male dolphins have been seen committing infanticide. When news of these nasty behaviors hit the press, *The New York Times* published an article entitled "Evidence Puts Dolphins in New Light, as Killers" subtitled "Smiling Mammals Possess Unexplained Darker Side." As with our earlier observations of male dolphins herding females to monopolize mating opportunities, the "dark side" of dolphin behavior always seems to come as a shock.

Personally, the discoveries we made about the behavior of male dolphins have enhanced rather than diminished my respect for them. They are more like us, and we are more like them, than some may want to admit. Their behavior is wondrous and surprising in its complexity, depth, and subtlety. They are not the perpetually sweet and kind creatures we at first imagined them to be. Dolphins can indeed be kind and sweet and gentle, but they can also be nasty and selfish and downright "politically incorrect" by our standards. Like us, they are complex and multifaceted.

MOTHERS,

DAUGHTERS,

AND SISTERS

While our research into male alliances was dramatic and shocking enough to attract a flurry of attention from the media, zoologists and academics, and dolphin and animal enthusiasts everywhere, a more subtle and more fundamental story, that of the lives of female dolphins, was gradually revealing itself to us, too. I say "fundamental" because it is generally the case with mammals that male behavior is shaped by the distribution and availability of females, while female behavior is shaped more directly and more intimately by ecological factors. The reason for this is that a female's success, in terms of leaving descendants, is determined largely by her ability to meet the energetic demands of pregnancy and rearing her offspring.

After the academic honchos Irv DeVore, Richard Wrangham, and Barb Smuts had visited Monkey Mia, Andrew and I began working toward our Ph.D. degrees at the University of Michigan. We left Santa Cruz, including my old VW bus, behind. Even if it had survived the drive from California to Michigan, it would not have made it through a brutal Michigan winter. It was a sad parting, but I like to think it is still roaming the sunny streets of Santa Cruz, hopefully with a new paint job.

Andrew had begun to explore the behavior of females and was conducting focal follows on a number of the females we saw regularly in Red Cliff Bay. Meanwhile Janet Mann, who was also a graduate student at the University of Michigan with Barb Smuts, joined our team with the intention of focusing on infant development. In time, many aspects of the lives of female dolphins in Shark Bay, which are, of course, intricately enmeshed with those of their infants, came into view.

It was April 16, 1990, when Nicky, who had grown from a rambunctious adolescent to a mature adult female of fifteen years, brought her second-born baby into Monkey Mia. Her first, born two years previously, had died, but today Holeyfin was a grandmother again. Like a cork set loose underwater, the newborn dolphin burst through the surface close against Nicky's side and took a gurgly breath. Awkward and wobbling on his newly unfolded fins, the baby inspired a gasp of delight from the gathered crowd of humans. This was his first day at Monkey Mia. Lightly pigmented creases, fetal folds, ran laterally up and down his sides, a reminder of the cramped quarters in Nicky's womb from which he had emerged sometime in the previous twenty-four hours.

Over the past few weeks, as Nicky lolled on her side in the shallows, looking overwhelmed by the magnitude of her belly, we had hoped and prayed for the opportunity to watch her give birth. We were disappointed but not surprised that we had missed it. Who knows where a dolphin would go and whom she might want as company during the process of childbirth? It must be a risky business given the numerous sharks that the bay is named for. But the birth remained her unshared secret, and all we could know was that somehow, somewhere nearby, she had succeeded.

If her birthing process was anything like those of captive dolphins, a few of which have been witnessed and even filmed, she would have gone through labor swimming slowly, pausing during contractions. The baby would have emerged tail first, not headfirst as with most mammals (probably to ensure the newborn doesn't attempt to breathe before being free of its mother's body). At the moment the baby came free, Nicky would have swirled around and quickly assisted her newborn to the surface for his or her first breath of air.

Now, the tension in Nicky's movements and the way the whites of her eyes showed revealed that she was feeling a bit stressed, perhaps hungry and concerned about her new role as a mother. She whistled over and over again, as if trying to imprint her whistle into the baby's mind.

For his part, the baby seemed to have no trouble keeping up with Nicky, even though she didn't bother to slow her pace much to accommodate him. In fact, Nicky was so eager for fish handouts that she was flailing about dangerously, and we were worried that she might knock into her newborn. But when the baby became separated from Nicky and swam inshore among the legs of his human admirers, she rushed in and pursued him down the beach, making a harsh, grating sound. Had he panicked or was Nicky trying to reprimand him? They returned a few minutes later, side by side; but with the buckets of fish empty, Nicky stayed a few yards offshore and calmed down.

I wanted to call this perfect, tiny dolphin Rabble, because I imagined him growing up to be a rambunctious rabble-rouser. He swam crisscrosses back and forth across Nicky's face, flopping over her rostrum and thrashing around a bit as if he hadn't quite got the hang of steering with his pectoral fins and pushing with his tail flukes. Nicky tolerated him, keeping one eye on the crowd. Holeyfin, Puck, and Surprise were nearby, milling around in the shallows, waiting to see if more fish handouts were forthcoming. Then, without warning, Nicky turned and charged at Surprise, screaming aggressively, open jawed, scraping her teeth against Surprise's tail stock. Immediately Puck and Holeyfin rushed to join and surfaced on either side of Surprise. Rabble, disoriented by the confusion, had become separated from Nicky and now found his way back to her side as she moved away from the other three females.

Surprise, a young adult female, probably just a year or two younger than Nicky, had been coming into Monkey Mia almost every day lately, taking fish handouts and playing seagrass games with people. She was a beautiful, sweet dolphin with a short rostrum and dark, perfect skin and eyes. She had always been friendly to our boats when we encountered her out in the bay and was a consummate bowrider. Her name was derived from a tendency to suddenly appear at the bow of our boat, seemingly out of nowhere. She had made occasional visits to Monkey Mia in years past and was particularly curious about people. Now she was becoming one of the regulars.

Nicky didn't seem pleased. We had seen her attack Surprise on several occasions recently. Perhaps she didn't like the idea of having to compete with yet another dolphin for her fish handouts, especially now with the added demands of pregnancy. Maybe she didn't want her baby to have too much exposure to the other females until he had thoroughly learned who his mom was. Or maybe she was just feeling irritable and

Surprise was her scapegoat. In any case, it was clear that their relationship was touchy. Holeyfin and Puck seemed to side with Surprise more often than not. I couldn't help thinking that perhaps this was just Nicky being crabby and unreasonable, while Holeyfin and Puck, more gracious in their acceptance of Surprise, were not willing to let Nicky's bad mood prevail.

Both Holeyfin and Puck had lost babies of their own within the past few months. Holeyfin, obviously very pregnant, had disappeared for a couple of days in January and then returned, thin, ravenously hungry, but with nothing to show. We would never know what had happened to the baby, if there had been a shark attack or the baby was stillborn or drowned. Nor would we know how Holeyfin felt about it. Puck had given birth to a healthy baby in March, which she brought into Monkey Mia immediately, but seven days later she showed up alone. The baby was gone forever.

Their stories were not unusual. Over the years we have come to learn that somewhere between one-third and one-half of the dolphin babies born don't survive their first couple of years. The causes of infant mortality are probably numerous. No doubt sharks are a nearly continuous worry, but there are other dangers as well. Some babies probably become separated from their mothers and ultimately end up stranding on shallow banks or drowning. Some may succumb to fatal encounters with poisonous stonefish, scorpionfish, stingrays, or other unfriendly creatures.

Having lost their babies, Puck and Holeyfin had both resumed cycling again and had been the objects of male attention for the past few weeks. They were spending a lot of time together. Both were without the energetic demands of pregnancy or nursing, and even old Holeyfin, usually stoic, reserved, and preoccupied with food, had been seen leaping around with pieces of seagrass in her mouth, playing keep-away with Puck, and frolicking with the adolescent males.

The contrast with Nicky, who had been grumpy, withdrawn, and feverishly obsessed with food, was stark and telling. Gone were the days when she spent most of her time cavorting around, a flirtatious tomboy, with her adolescent buddies: Snubnose, Bibi, Gamma, Wave, Shave, Lucky, Pointer, and Lodent. I felt a touch of remorse, watching this transition. Her easy, playful nature was suddenly muted. She was an adult now, no longer practicing for adulthood and no longer subject to the permissiveness granted to youth. She would have to play by the rules, no matter how harsh. Her relationships with Wave. Shave, Snubnose, Bibi, and Sicklefin would be forever altered by the tensions, limitations, and imperatives of adult sexuality and politics. With the demands of motherhood, she would need to remain focused on the substantial task of getting enough to eat for both herself and Rabble.

I can relate to Nicky, having made a similar transition recently myself. A few years ago I entertained notions about various different men who attracted me and enjoyed few pleasures as much as a good flirtation. I traveled back and forth to Australia to watch dolphins and never let anything, including those notions and flirtations, interfere much with my plans to succeed with my work, to have fun, and to do pretty well as I pleased.

As my biological clock ticked away, my attitude suddenly began to shift. I began to think differently about what I wanted from a partner. Then I met a man (now my husband) who was just right. In no time we married and I became pregnant. So ended my days of carefree travel, thick with possibility and sprinkled with romance. Even my research trips to Australia have fallen by the wayside lately. I suppose some of my old friends find me less fun these days, and certainly my relationships with other men have changed forever, but my priority these days is ensuring that my family is safe, healthy, happy, and well fed. In spite of all these changes and some chronic sleep deprivation, I find more pleasure in my newfound family life than I ever imagined possible.

For a female dolphin, getting enough to eat is basically a full-time job. Fish are nutritious, full of protein and fat, but they can also be hard to find and to catch. For some females, the demands can become simply unbearable.

Yan, a large female we have known from offshore since the beginning, in 1991 gave birth to a baby whom we dubbed "Yoda." When we first saw them that year, Yoda was a few months old already, and Yan looked okay, perhaps a bit thin. But as our field season progressed, she looked worse and worse. First she developed a sunken-in look around her back and tail stock, then around her neck. Later, her ribs became visible and her breath noticeably rank.

Yoda too became emaciated and, unlike other babies, spent all his time right alongside Yan, as if stuck to her side by a Velcro patch. Yan foraged with an intensity of focus that was impressive. Always on the move, always seeking, always working hard. When we encountered her, she graciously approached the boat for a brief greeting, then went back to her business. In the end, she couldn't make ends meet. Perhaps her milk supply dwindled, perhaps she was not well to begin with. One day later that year, we saw Yan without Yoda. Yoda was dead. Over the following months, Yan recovered, gaining back all of the weight she had lost.

When their babies survive, female dolphins nurse for at least about four years and then start to cycle again. Their hormones shift, they begin to ovulate, and they go through periods where they are tremendously attractive to males. If their babies don't make it, as in the case of Puck and Holeyfin, they resume cycling shortly after losing the infant. Once again they are subjected to being chased, charged, knocked at, forced to stay with, and perhaps even forced to mate with those males who take an interest.

It is difficult to tell how females feel about being herded. I can easily project my own opinions and imagine that these gals are furious, rebellious, and terribly sophisticated in their monkey-wrenching techniques, but more honestly, I would have to say that such emotion is not obvious.

In some animal species, females quite effectively choose with whom they mate. They may choose males likely to pass along valuable traits to their offspring, such as health, vigor, size, or cleverness. In species where males contribute to parental care, females may choose males who show signs of being good providers.

Female dolphins, as in most mammal species, probably have some opinions about whom they would prefer to mate with. Males don't contribute to child care, so females probably aren't looking for paternal males, but they may prefer males who are large, healthy, smart, good hunters, good echolocators, or whatever else might make a male dolphin attractive to a female dolphin.

Faced with at least two or more males working together, however, a female dolphin probably has little opportunity to put her preferences into practice. Male dolphins can be so aggressive in their herding of a female that she may even be forced to mate against her will.

Being herded by males can be a troublesome, even downright dangerous ordeal for a female dolphin and even for her offspring. Males usually prefer to herd females who could become pregnant (not females who are already pregnant or have newborns, with a few exceptions). By the time a youngster is about four years old, his mother will begin to go through estrous cycles again and become attractive to males. Youngsters still spend much time with their mothers at this age, so they are often party to the herding of their mothers. One female we called Munch even lost her youngster while she was being herded. Snubnose,

Bibi, and Sicklefin had forced her to remain with them in the shallows at Monkey Mia for the better part of several days. They were particularly brutal to her: constant knocking repeatedly escalated to aggressive head jerks and charging, screaming, chasing, and hitting. Munch was obviously distraught and uneasy about being close to people. Her baby, who was fairly small and thin for his age, was apparently so afraid that he remained offshore, apart from her. I could hear him calling to her incessantly and caught occasional glimpses of his little dorsal fin as he surfaced offshore by himself. But Munch was unable to go out to him, and after that we never saw him again.

Another female we called Chiclet became a favorite target for Snubnose, Bibi, and Sicklefin in 1987 and 1988. When she was first brought in to Monkey Mia, she looked horrible—her ribs showed, and she was covered with scars and sores and bumps. There was a certain desperation evident in the way she moved, always fast and jerky and wide-eyed. We took pity on her, as it seemed unlikely that she could tolerate being held hostage at Monkey Mia, where she would be unable to forage, for days on end. So we took to offering her fish handouts. When I fed her, she awkwardly but effectively gobbled what was offered with no hesitation whatsoever, then turned to my hand with jaws agape, pushing against me, bumping me insistently with her rostrum. She was obviously hungry enough to overcome any fears she might have had and eagerly sought more food. We made a point of feeding her well that day and on subsequent occasions when she was herded into the shallows at Monkey Mia.

Holeyfin herself suffered a horrible injury while being herded by Snubnose, Bibi, and Sicklefin. It happened during a time when I was back in the United States. Nobody knows for sure what happened, but one morning they all failed to show up in the shallows of Monkey Mia as was their usual pattern. The males had been herding Holeyfin, who was still accompanied by four-year-old Holly. Later, when they finally did show up, they were a gruesome sight. All had burns, apparently where their bodies had been exposed to the drying and burning effects of the sun. We can only guess that the males had been chasing Holeyfin around when all had become stranded on a sand bank. Holeyfin was by far the worst. About a third of her back and shoulders had deeply blistered, and the skin had come off. What remained oozed a thick gooey whitish substance flecked with blood and exposed to the salty sea. She had apparently been stranded farthest out of water. Holly had a stripe of burned skin down the middle of her back. She had probably been alongside Holeyfin but, being smaller, had been less exposed. The males had thin strips of burned skin where only a small part of their backs had been exposed. Apparently they had been behind and in deeper water. It would take many months for Holeyfin to recover from these burns, and the prominent scars stayed with her for the rest of her life.

Sometimes females put up quite a fight when they were herded, apparently determined to escape their suitors. Other times they seemed resigned to their fate, nonchalant about the whole matter. The males might be particularly vicious, with nonstop knocking, chasing, hitting, and screaming on one occasion, and on another they might simply follow the focus of their attraction around as she foraged. There might be a little skirmish, some knocking and chasing, but for the most part, the males were reasonable and the female free to go about her business.

We really have little understanding of the reasons for this variation. It could reflect differences in the stage of the female's cycle and therefore her interest in mating or differences in the preferences of females—that is, whether she wants to mate with these particular males. The Monkey Mia males, Snubnose, Bibi, and Sicklefin, were consistently very aggressive herders. Females herded into the shallows of Monkey Mia were especially eager to escape. Not only were they forced into proximity with humans, but they could not forage for themselves while in the shallows. The males were eating their fill of handouts while the herded female, already stressed, went hungry. One interesting observation suggests that the males might even have been aware of this dynamic.

It was March of 1988, and the males were herding Jag. Bibi had been especially keen, remaining close to her and knocking fiercely whenever she strayed too far. Snubnose and Sicklefin as well as Holeyfin, Puck, and Nicky were all being fed buckets of fish by the tourists. Bibi dashed in occasionally to grab a fish, then returned quickly to his position near Jag, a bit offshore of the feeding activity. Then he did a very surprising thing. He dashed in, grabbed a fish, and returned to Jag, still holding the fish in his mouth. He was whistling and making a series of high-pitched, squeaky sounds we referred to as "food calls" (the dolphins almost always made these sounds when offered a fish). Bibi mouthed the fish, and I thought he would swallow it, but instead he dropped it, then even pushed it with his rostrum toward Jag, who grabbed it and ate it. He was feeding her! Was he aware of the fact that she was hungry? Was he trying to win her favor? It was a surprising act of kindness in the midst of what otherwise looked like a rather mean business.

Faced with the overpowering force of cooperating males, why don't females counteract the males by forming their own alliances? This is a central and perplexing question. There have been a few, albeit sparse, hints suggesting that females do occasionally help each other out when faced with marauding males.

One day, Snubnose, Bibi, and Sicklefin were herding Poindexter (whom we met in the previous chapter) at Monkey Mia. Nicky, Puck, and Holeyfin were also around. We waded into knee-deep water to watch them and record their vocalizations. When buckets of fish arrived, the males were distracted, and Poindexter took advantage of the confusion to bolt. A trail of fluke prints on the water, leading straight out to the north, was all that could be seen of her departure.

Several seconds later, Snubnose turned to look and discovered her absence. He panicked, and soon so did Bibi and Sicklefin. All three of them began racing back and forth, searching frantically for Poindexter. Having determined that she was nowhere nearby, they went leaping off to the north, the direction in which she had gone a full minute or so earlier. We could see them still leaping back and forth half a mile offshore. They obviously had not found her but were not going to give up.

Then a surprising thing happened. Poindexter returned to the Monkey Mia shallows on her own. She was a completely wild offshore dolphin and thus never came into Monkey Mia voluntarily. But here she was. She swam directly into the shallows, where she was joined by Nicky and Puck, who took up position alongside her. The three females, their bodies pressed close together as if holding hands, quietly slipped away out to the northeast. Poindexter was rarely seen out east of Monkey Mia; rather, she was usually seen out to the northwest. She was heading off in a direction where the males would not expect her to go. Nicky and Puck were escorting her through the narrow channel of deeper water that was the only way to get through when the tide was this low. A while later the males returned to the Monkey Mia shallows, having given up on their search.

Once Richard observed a male alliance separate a female (never identified) from her group and pursue her for some distance when suddenly another group of six dolphins in turn began pursuing the males. The six were all females, and when they caught up to the males, the males gave up their pursuit of the female they had been chasing. Perhaps they were intimidated.

Andrew saw females on numerous occasions "pairing"—that is, swimming in synchrony with one female in a position alongside and slightly to the rear of the other, with her pectoral fin pressed against the forward dolphin's side. In some cases, pairing happened in situations where the females were being harassed by males. Once two females paired when some males joined their group, then split up again after the males left. Once a female interposed herself between a group of males and the female they were harassing and paired with her. She was later chased off by the harassing males. Another time two pairing females simply departed from the company of some males who had been harassing one member of the pair. These observations suggest that males are somewhat intimidated by pairing females, but they are hardly convincing evidence of female cooperation to thwart male harassment. Especially since females often paired in situations where there were no obvious correlates with male behavior.

The alliances of male dolphins are so dramatic, one can't help but notice them. They spend virtually all their time together and engage in all sorts of petting and social interaction to maintain amicable relations. Females just don't seem to have much interest in cooperating with one another. It may simply be that they are too preoccupied with foraging or that opportunities to cooperate effectively enough to actually thwart male harassment are just too rare or dangerous.

Barb Smuts, my adviser at the University of Michigan, studied baboons and chimpanzees before she joined the dolphin team. Reviewing information from species after species of mammals, she found a common pattern. Males, singly or in cohoots with other males, often use aggression to get females to have sex with them. In some cases males may actually force females to mate, but often they rely on threats and attacks. In some monkey and ape species, females sometimes cooperate to thwart male aggression, but such female cooperation is basically nonexistent in other mammals.

This is mysterious, and I write it with some sense of horror and regret, since it goes against any feminist notions I harbor about how the world should work. But there is no denying it. Although it seems they could benefit from helping each other, female dolphins just don't do it. I want to stand up and yell out, "Get together, sisters, and take a stand! United, you are powerful." But they are dolphins and I am human, and I cannot expect them to live by my moral standards.

Even though they don't regularly come to each other's aid, all female dolphins in Shark Bay spend most of their social time with other females. Some have long-lasting "friendships" and spend a lot of time in the company of certain other females. In at least some cases, we know these friendships involve mothers, daughters, and sisters. Other females spend time with whoever happens to be around, the social butterflies who seem to get along with anyone and everyone equally. Then there are a few loners who rarely associate with anyone else other than their own offspring. Maybe they have lost or moved away from kin? I feel a bit sorry for them. Being a female dolphin is hard enough without facing it alone. My impression is that for a female dolphin, the greatest pleasure is to be a matriarch, the central figure in a large and healthy family. But that seems to be an extremely difficult goal to achieve.

It is September 1992, the start of another birthing and mating season. Holeyfin is attracting the males' attention once again. In spite of her worn-down teeth, her missing tail fluke, and the sunburn scars on her back, not to mention what strikes me as a touch of dolphin senility, she still ignites a fire in the hearts of many of the Shark Bay males. She is the dolphin equivalent to Jane Goodall's ugly old female chimpanzee Flo, whose very active and romantic (for a chimpanzee) sex life seemed only to intensify with old age.

For the past several days Holeyfin has not made her usual appearances at Monkey Mia. Though at first her absence sent the usual wave of concern through camp, we were not too surprised to find her in the company of Realnotch, Hi, and Bottomhook. They followed her closely, and once I heard Realnotch knocking at her as he was snagging near the surface. She seemed okay, foraging in the shallows off Cape Rose alongside them, though I wondered if she really knew how, given all the time she spent in at Monkey Mia getting easy handouts.

I feel a certain sense of futility, watching Holeyfin go through the motions of being herded once again. Nicky and Joy are her only surviving offspring. Since she lost Holly, she had gone through three pregnancies, all resulting ultimately in naught. It seems unlikely that she will pull it off this time given her age and her past performance.

To top it off, she lost her second grandson, Nicky's son, Rabble. He showed every sign of being healthy and had taken quite a shine to interacting with people at Monkey Mia. One day he arrived at Monkey Mia with a gruesome wound, several parallel lines where the flesh had been rent by the razor-sharp teeth of a shark. We thought he would die then, but he managed to recover. His demise came about later, around the time he was weaned. He became completely dependent on handouts from tourists and spent very little time offshore. We worried that he was not learning to be a "real" dolphin, to hunt for himself, and was not developing relationships with other dolphins. Then he failed to show up at Monkey Mia and was never seen again. One more statistic in our records, one more tragedy in the Holeyfin matriline.

WHISTLES

AND CLICKS

Holeyfin was herded by several different male alliances during the summer of 1992. Then, a year later, she gave birth to Hobbit, another male infant. All went well at first, until one day in January of 1994. Holeyfin brought Hobbit into the shallows at Monkey Mia as usual. But then, as she was busy getting fish handouts from the tourists, leaving Hobbit on his own a bit farther offshore, an enormous shark slid in from deeper waters and grabbed Hobbit. Horrified tourists watched as the water turned red with blood and the shark circled back to finish its gruesome business.

By the time Holeyfin got to the scene, it was too late. She had lost yet another baby, her fifth loss in a row.

Holeyfin and the other Monkey Mia dolphins were particularly unlucky, losing most of their infants to one thing or another. But even the babies born to offshore mothers died at an alarming rate. Usually they just disappeared, leaving no trace as to the cause of death.

Not too unlike humans, dolphin babies nurse many times each day and are not weaned until they are at least four years old. During the first two weeks or so after birth, babies stay right alongside their mothers virtually all the time. Over the next few years they continue to spend a lot of time with their mothers, benefiting not only from having a safe side to swim at, but also from the opportunity to watch and learn from their mothers the skills they will need in order to survive. Yet, after following mother-infant pairs for some time, Janet and Barb had discovered something quite puzzling: When babies are just a couple of weeks old, they begin to wander off, and neither they nor their mothers seem much concerned, even when the baby is one hundred or even three hundred yards away. What human mother would allow her newborn infant to stray off unattended? It seemed awfully risky. Presumably dolphin moms would provide some measure of protection to their infants if they were nearby, but at several hundred yards apart, in the event of a shark attack or any other dangerous situation, help could easily arrive too late, as had happened with Holeyfin and Hobbit.

There must be some major advantage to separations, either for babies, mothers, or both. Since babies spend most of their time socializing with other babies while they are apart from their mothers, Janet and Barb reasoned that it might be critical for babies to socialize with other babies, developing the relationships that will be central to their future as adults.

Another factor is the ecology and energetics of motherhood.

Mothers need every advantage in order to acquire enough calories to successfully rear their infants. Foraging with a baby in tow might be so burdensome that it is worthwhile to trade off for the risks of separation. If a mother obtains enough food to produce plenty of milk, her infant can grow more quickly, thereby becoming less vulnerable to shark attacks sooner. If a mom keeps her baby in tow, she may be better able to protect him but not able to provide enough milk to support the baby's rapid growth, so her baby remains vulnerable for longer. Trade-offs.

Whatever the advantages and disadvantages, the fact of these perplexing mother-infant separations remains. Since they cannot see each other at a distance of more than a few feet, we assumed that mothers and infants must have some other effective way to keep track of each other's whereabouts so that they could quickly and easily navigate back and forth. We guessed that they must use some sort of vocal signal.

Seals, bats, birds, reindeer, mothers, and babies of many different sorts have individually distinctive calls that they use to keep in touch with each other. This is especially true for animals that live in colonies and must separate and reunite within the context of many other individuals. Some species of bats, for example, live in colonies where several hundred individuals all share a cave wall. Mothers leave their babies parked in the crowd to go off foraging, then must locate them when they return. It is in the interests of both the mothers and the babies that they be able to find each other. The same is true for some seals that breed and give birth on land in big colonies. Mothers go off to sea to hunt and feed, leaving their babies behind in the crowd. Later they have to find each other. What better way than to have mother, baby, or both, for that matter, make some distinctive vocalization that will let them find each other in the crowd?

Baby dolphins in Shark Bay must be faced with a similar situation. Although they don't live in colonies in quite the same way as bats or seals, in effect the potential for confusion is probably similar. Sound can travel far and fast underwater. At any time, a dolphin may be able to hear the sounds made by many other dolphins, even if they are not in the immediate vicinity. It must be critically important that mothers and babies be able to recognize each other in that sea of sounds and, given the dangers they are faced with, to do so quickly and effectively.

Dolphins make a huge variety of different sorts of sounds, but we decided the most likely candidate for mothers and infants to use to identify each other was a whistle. Whistles are tonal sounds, usually around one to three seconds long. During that time, the frequency (pitch) of the whistle usually changes. A whistle may rise in frequency and then fall again, or it might start high, descend, and then rise again or start with a rise and then level off. There is effectively an infinite variety of ways in which the frequency can be modulated, resulting in different whistle contours.

David and Melba Caldwell were among the first to take on the challenge of studying dolphin whistles. This husband-and-wife team spent many years watching and recording dolphins in various oceanaria. Recording dolphin sounds is fraught with difficulties, not the least of which is the problem of determining which dolphin in a group was responsible for producing a given sound. Dolphins produce all of their vocalizations internally, with no visible moving mouthparts that an observer can use to figure out who is making which sounds. When a dog barks, he opens his mouth rhythmically at the same time. When a human speaks, her lips and tongue move around. With dolphins, since all sounds are produced internally, there are no clues visible to an outside observer to indicate who said what.

The Caldwells skirted around this problem by recording dolphins as they were held stranded out of water for veterinary care procedures. In air, Melba and David could hear the dolphins' whistles, be sure about where they were coming from, and record them with a microphone held right up to the dolphin's forehead.

What they found was this: Each dolphin had a unique whistle. Some dolphins made more than one type of whistle, but all of them had one unique type that accounted for most of his or her whistles. The Caldwells called these dolphin-specific whistles "signature whistles." They suggested that dolphins use their signature whistles to keep track of each other's whereabouts by announcing their presence to other dolphins: "Hey, Puck over here"; "Nicky over here"; "Swims-with-Sharks out here"; "Mooncraters down here."

The signature hypothesis makes a lot of sense from a dolphin's perspective. Dolphins can't see each other unless they are right next to each other, and they have very strong preferences regarding the company they keep. If dolphins didn't care about this, a universal "Dolphin over here" signal would suffice. However, for other dolphins to know for certain that they hear Puck, for instance, Puck would have to produce a "Puck here" signal. When another dolphin heard Puck's whistle, he or she would need to be able to tell that whistle from others and also to associate it with the dolphin Puck.

What is especially interesting about signature whistles is that baby dolphins aren't born with their signatures, nor do they inherit the whistle of their mothers or fathers. They make gurgly, irregular, messy-sounding whistles at first. Gradually they refine their sound into a clean, unique whistle by the time they are about six months to a year old.

This may seem unremarkable to us, since we are accustomed to learning the complexities of human speech, but in fact we are quite exceptional. Few animals other than humans and songbirds learn the sounds they use to communicate with as adults. Even among our primate relatives, youngsters don't seem to depend much on learning to produce the proper sound types as adults. For example, young monkeys raised in isolation still make the sound types appropriate to their species. They may not use them appropriately, but they can produce them.

The fact that dolphin signature whistles *are learned* suggests that there must be something unusual about dolphin communication and raises a series of fascinating questions: What is it about the dolphin way of life that would require that they learn their whistles? Just how do they learn their signatures? Do their mothers give them "names"? Do they simply invent a whistle as infants? Do they adopt the whistle of another dolphin? Do they keep the same signature for life?

Dolphins have another remarkable capability: They are excellent vocal mimics. Trainers working with dolphins in marine parks use whistles to instruct their pupils, just as dog trainers do. Much to their surprise, a few trainers discovered that the dolphins were whistling training signals right back at them. They were imitating the trainers' whistles.

Sam Ridgway, a renowned dolphin veterinarian, once had some white whales in his charge (white whales are related to bottlenose dolphins). He and the trainers kept hearing what sounded like voices in the distance, a barely discernible conversation. They assumed for some time that it was people talking over on a pier, some seventy-five meters away. Then they discovered that the sounds were actually coming from one of the whales, imitating the pitch and rhythm of human conversation.

In one study, dolphins were trained specifically to imitate a variety of computer-generated whistles. In some cases the dolphins immediately produced accurate renditions of the same contour. Even when the computer-generated whistles were quite bizarre, the dolphins made discernible attempts to imitate them.

Once again, to a human, this might seem unremarkable because we are so good at imitating sounds ourselves. Vocal imitation is a large part of how we learn to speak. But very few animals, with the exception of a few bird species, seem to be able to imitate sounds. Parrots and mynah birds can learn to imitate human speech, and mockingbirds, starlings, and some others will imitate sounds and songs of other birds.

Dolphin vocal learning and imitation skills are probably not unrelated abilities, but rather indications of a generally flexible and sophisticated communication system. One possible reason for such flexibility was raised by Peter Tyack. He recorded whistles from two captive dolphins, tank-mates named Scotty and Spray. He discovered that each dolphin had a favored whistle (that dolphins signature) and a secondary whistle. The catch was that Scotty's secondary whistle was Spray's signature, and vice versa. Apparently each dolphin frequently imitated the other's signature. Peter reasoned that the dolphins were doing this to establish contact, essentially saying, "Hey, Scotty, are you there?" instead of just, "Spray here, Spray here, Spray here." The use of signatures in this way has obvious parallels with human use of names. And because it requires a dolphin both to develop a unique signature and to be capable of imitating other signatures, it could at least partly explain the need for vocal learning and imitation in dolphins.

Until quite recently, the discoveries about dolphin signature whistles, learning, and imitation all came from studies of dolphins held in captivity (and they probably would never have been possible otherwise). But this raised one question: Is this "normal" for dolphins? After all, a dolphin that has been in captivity for some time or is being stranded out of water and handled by veterinarians might very well be acting a bit strangely. Moreover, even if the dolphins weren't stranded and handled, life in a tank certainly seems to have the potential to dull their vocabulary and their wits.

To understand how dolphins communicate with each other under natural conditions would require watching and recording a population of recognizable wild dolphins. The Shark Bay dolphins seemed like a good bet, and the conundrum of mother-infant separations provided us with a lead that might prove fruitful in discovering something about how whistles are used in the wild. But there were some major logistical problems to overcome.

In 1990, while doing coursework at the University of Michigan, Janet and Barb and I teamed up and devised a plan to examine the ways in which dolphin mothers and infants keep in touch during separations. We hired Julie Gros Louis, all big brown eyes under an astonishing head of corkscrew curly hair, to be our amiable and hardworking research assistant, and in April of that year Janet, Julie, and I set off for Monkey Mia to follow and record sounds from mothers and babies.

Cookie was two years old. His mother, Crookedfin, was one of the regulars at the Monkey Mia beach and, conveniently for us, spent a lot of time in the waters just offshore. Cookie, a sleek, dark dolphin with a pointy, swept-back dorsal fin, was best friends with Smoky, another two-year-old. Smoky, a stocky, soft-gray dolphin, had a blunt-topped fin with a distinctive wave in the trailing edge. His mother, Yogi, also frequented the waters not far offshore of Monkey Mia.

Cookie and Smoky took advantage of every opportunity to seek each other out and play, splashing and chasing, goosing and jawing at one another, charging and leaping over each other, stroking one another with their pectoral fins, playing keep-away with bits of seagrass, and occasionally breaking off their play to chase little fish. Both were overflowing with that exuberant, carefree little-boy energy that thrives on freedom and fun, and they were nearly always together, except when they got hungry or tired, at which point they would head off in search of their mothers.

By watching Cookie, Smoky, and some of the other babies, we hoped to learn how they found each other when they were ready to play, then found their mothers again when they were ready to "go home."

There was a good chance of knowing for sure the exact source of the whistles we recorded since they were often alone as they traveled back and forth; but we also needed to be able to listen and record continuously—a problem, given that dolphins are almost always on the move. In order to keep up with them, we would need a motor, but motors make a lot of noise, drowning out the dolphin sounds. We were in a bind. Either we could hear or we could see, but not both.

In the past, to record dolphins offshore, we would position ourselves alongside them, turn off the motor, drop the hydrophone into the water, and record until the dolphins moved too far away for us to see what they were doing, which often happened fairly quickly. Then we would stop recording, pull up the hydrophone, cover everything to protect it from spray while we were under way, start up the noisy motor, zoom to catch up and reposition ourselves alongside the dolphins again, put the hydrophone back into the water, and start recording again. We were so tangled up in and distracted by our equipment that there was little opportunity to actually watch the dolphins. We needed a way to move along with the dolphins and hear and record their sounds at the same time.

We borrowed an idea from Laela Sayigh and Peter Tyack, dolphin researchers from Woods Hole who had been struggling with the same problems in their studies of dolphins in Florida. They had worked out a system in which they powered their boat using an electric motor, which is much quieter underwater. They towed a hydrophone alongside, attaching lead weights to the cable to keep it from trailing up to the surface, where it would bounce along and make a deafening racket. Although it was by no means a perfect system, we were optimistic and arrived at Monkey Mia carrying the most powerful electric motor we could find and a new hydrophone in our luggage.

One day in early April, Crookedfin, with Cookie in baby position, approached the shallows at Monkey Mia. But as soon as they got close,

Cookie broke away, turned his back to the shoreline and the people, and snagged, looking dejected. Underwater I could hear him whistling incessantly. His distinctive whistle rises fast and high and then descends more slowly. He seemed to be whining to his mother. I could easily imagine him saying, "C'mon, I don't like it here, I wanna go, let's go *now*, I'm hungry, c'mon, Mom, *c'mon.*"

Other babies, like Holly, for example, have taken a shine to people right from the start, but not Cookie. Maybe he had a bad experience early on, or maybe he finds us boring. After watching his social life offshore with Smoky and the other dolphins, I can imagine that a line of people, all kneecaps and feet underwater and from above, hands wiggling, grabbing, reaching, and touching him, trying to tempt him into taking dead fish handouts day after day, would not rate as a good time.

Floating just offshore with his back to the people, I could sense his disgust. One hundred feet farther offshore, Surprise surfaced and headed toward us. As she got closer, Cookie kicked his tail flukes and approached her. Just as he reached her, he tilted his belly toward her and dove directly underneath her. She dove also, and the two surfaced a moment later, wiggling and splashing. Cookie rolled belly up at the surface, revealing a pink and erect penis. He and Surprise rolled around near the surface. Parts of bodies poked out through the water surface and wiggled and splashed around. At one point Surprise chased Cookie, opening her jaws against his side, jawing at him, perhaps scraping him with her teeth.

They drifted farther and farther from shore until they were almost two hundred yards from Crookedfin and the shallow water of the Monkey Mia feeding area. I could just make out where they were, but not much of what they are doing. At the shoreline, Crookedfin was still pacing up and down the line of people, occasionally stopping in front of someone, mouth agape, asking for a fish. She had been fed three large mullet and five bony herring all in the space of half an hour. Easy pickings. I heard a few squeaks and growls and grates and squeals coming from Cookie and Surprise offshore and echolocation clicks coming from Crookedfin. Then, abruptly, the sounds from offshore ceased, and I heard Cookie's whistles. They were a bit faint, but nonetheless distinctly his whistle, over and over again. I saw him surface briefly, now fifty feet closer, having left Surprise farther out. The whistles grew louder and louder as he approached, making a beeline toward Crookedfin (and me). As he got within about ten feet of Crookedfin and the line of humans, he stopped, turned around, snagged for a minute, then settled with his back to the people and Crookedfin. Waiting. He had stopped whistling.

Cookie whistled a total of eighteen times, all as he was traveling back to join Crookedfin. Was he whistling to inform Crookedfin that he was on his way home ("Here I come") or to ask her where she was ("Are you there?")? I had expected that Crookedfin might whistle back to him and so had positioned myself right alongside her as Cookie was coming in, but she was silent. In fact, she showed no signs of interest in Cookie whatsoever. Did she assume that Cookie knew where she was? Why not at least answer him?

A few minutes later Crookedfin headed offshore, passing by Cookie, who swung into baby position alongside her. The pair traveled back out to the northwest. We loaded up the boat and headed offshore to follow them. Cookie was still in baby position when we caught up, and he seemed sleepy. Perhaps all that tumbling around with Surprise had worn him out. He remained tucked up close against Crookedfin's side, moving with her as though in his half-conscious state he had relinquished control over his body to her. But a few minutes later he abruptly perked up and left Crookedfin's side, heading away from her as if on a mission. Crookedfin did a tail-out dive, back to foraging. We followed Cookie.

Moving under power of our electric motor, we found it hard to

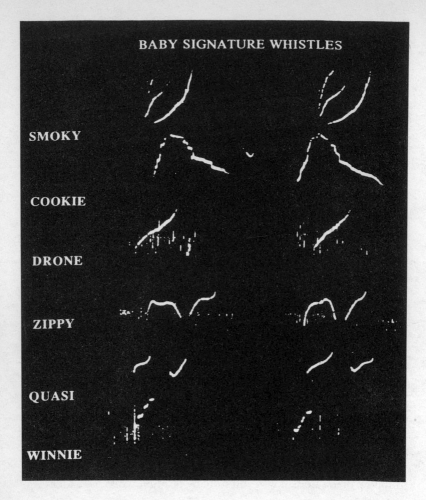

BABY SIGNATURE WHISTLES

SMOKY

COOKIE

DRONE

ZIPPY

QUASI

WINNIE

keep up with him. We didn't hear any whistles, but there was a lot of noise caused by the hydrophone dragging through the water. Cookie traveled about fifty yards before we noticed another tiny fin approaching: Smoky. The two dolphins sped up as they got closer, then raced together, splashing, tilting, and diving as they met. Playtime! They surfaced a moment later side by side, Cookie stroking Smoky's side with his pectoral fin. Smoky first rolled sideways so Cookie's fin would stroke his belly instead, then slid back so Cookie's fin was stroking him under the chin. They disappeared from view again and left us waiting up above.

After watching them play together for a while, we heard a whistle that sounded just like the plaintive tones we had heard when Cookie was coming into shore earlier. Then another whistle, different from Cookie's. It was Smoky's short, steep, rising whistle. They both whistled a few more times, and then we saw them surface, about forty feet apart from one another and oriented in opposite directions. They both headed off toward their mothers.

We followed Smoky this time. He continued to whistle regularly, picking up speed, and was soon traveling rapidly in a direct line, presumably toward Yogi. All we could see was an occasional dorsal fin popping up as Smoky took a quick breath and then disappeared underwater again. Thirty yards farther on, he popped up again for a breath. Farther out in the direction he was traveling, we finally caught a glimpse of what we presumed to be Yogi. When we caught up to her, Smoky was already in baby position alongside her. We settled in, watching mother and baby surface and dive together, certain that before long, Smoky would head back over to Cookie, wherever he was, for another play session.

We had expected mothers to answer their babies and wondered if they would answer with their own signature or by imitating their baby's signature. But it looked as if our earlier experience with Crookedfin at Monkey Mia were typical. We heard very little from moms and so were puzzled as to how babies knew where their mothers were. Do they just take a guess that she is in roughly the same place as before? I imagine the underwater world as relatively featureless, and the task of navigating back to the same place must be challenging for a baby dolphin. Babies may hear the echolocation clicks produced by their foraging mothers, but it seems odd that a mother wouldn't take some small precaution to help out by clearly announcing her location. We usually followed babies rather than moms, so there was a chance that we simply weren't hearing the mothers' whistles because we were farther away from them. But there were times when we were close enough and the recording conditions good enough that we *should* have heard mothers if they were in fact whistling.

Still, we decided to try following moms instead of babies. One day we followed Yogi as she headed away from where Smoky was playing with Cookie. She was foraging intently. Smoky had been gone for about forty minutes when Yogi stopped foraging and began to travel quickly, still headed away from Smoky. We pulled the boat up ahead of her and sat stationary, listening for whistles as she came toward us. Suddenly she turned about-face, whistled three times, and then headed back in the opposite direction, toward Smoky. She didn't whistle repeatedly, but there was no mistaking it: she had whistled just as she changed direction and headed back toward Smoky.

So mothers do whistle to their babies, at least some of the time, but as far as we could tell, they do so with just one or two whistles—easier to miss—instead of a steady stream of repetitions. There was no way Janet and I could follow both mothers and babies at the same time, so any further understanding of the mother's side of the story would have to wait.

Late in August, on our final day at Monkey Mia for the year, Janet and Julie and I watched Cookie and Smoky chasing tiny fish together. Both turned belly up just under the water surface and zoomed around, buzzing like a pair of electric shaving devices gone haywire, twisting and turning as the fish evaded them. On a few occasions they almost collided with each other as both pursued the same fish. Later they both went down to the bottom in shallow water over a seagrass bed. Both poked their beaks into the seagrass and made the peculiar sounds we had often heard other dolphins make when they were doing the same thing. Smoky suddenly darted off and poked his beak into another clump, and Cookie followed. They were bottom grubbing together for fish that hide in the seagrass. They broke off occasionally to tumble around, charging gleefully together into a swirl of churning and splashing. Several times they returned to their mothers for some "R&R," whistling as they went.

We had spent most of our days over the past few months watching these two babies along with some of the others, Quasi, Winnie, and Peglet. Each day of their young lives was so packed full of fun and adventure, it was hard to accept that we would have to leave them now and miss out on the next several months. Even more heartbreaking was the thought that they, like all baby dolphins in Shark Bay, faced a high probability of dying before they reached adulthood. We would be back next year, and hopefully they would be here so that we could follow their progress.

Although we had probably raised more questions than we had answered, we now had good evidence that Shark Bay dolphins *do* have signature whistles. We had even been able to watch some of the details of how babies used their signatures in the process of navigating back and forth between their mothers and their playmates. Virtually all of the studies of dolphin whistles prior to this had relied on capturing and recording dolphins held out of water; this was the first time anyone had been able to watch dolphins actually using their whistles under natural circumstances in the wild.

Meanwhile, and for several years previous to and following our field season with Cookie and Smoky, practically every morning had been devoted to recording whistles and other sounds from the dolphins visiting in the shallows at Monkey Mia. Andrew and I had worked out a good system for recording that allowed us to identify which dolphin made a particular whistle. One of us, "the recorder," would hold the tape recorder in a backpack while standing in the water and dangling a hydrophone. The hydrophone was connected to one channel of the tape recorder, while a regular, in-air microphone was connected to the second channel. The person acting as the recorder had to be still, otherwise there was nothing to hear but the crunch and grind of sand underfoot.

The other person, "the identifier," was free to walk around and would try to stay as close to a dolphin as possible in order to hear his or her whistles, which were audible in air as long as the dolphin's head was at the water surface. When a dolphin whistled, the identifier would point out the dolphin and the recorder could speak the whistler's name onto the second channel of the tape recorder.

We had been watching Snubnose, Bibi, and Sicklefin almost every day, and we probably knew more about them than any other dolphins in the bay. Just from listening to their whistles, I could tell that they didn't have distinct signatures the way the babies had. Something funny was going on with them. When we got back to Michigan, I began the process of analyzing the whistles we had recorded from the three males over a period of several years.

The first thing I did was to enlist the aid of my friend and colleague John Pepper. John had visited with us at Monkey Mia (and met his wife, Gillian, on the plane ride over) and therefore had some familiarity with the dolphins. Brilliant, but humble to a fault, John is also one of those lucky souls gifted with a knack for numbers and computers. We spent literally hundreds of hours sorting through the 250 tapes, painstakingly listening for identified whistles, then digitizing the whistles onto a computer, sorting them by dolphin and by date, printing out spectrographs of each one, and, finally, trying to figure out what it all meant. This one small paragraph summarizes what took a phenomenal amount of work over several years to accomplish.

In the early years of our recording effort, 1984 and 1985, the males had been making many different kinds of whistles. They didn't seem to have signature whistles at all. Some of the whistles we recorded from them were obviously different types, with contours that were repeated from time to time and looked somehow deliberate in form to me. But many others were sloppy and irregular. They seemed to be haphazard whistle noise, as if the dolphin really didn't have any particular whistle in mind. This didn't fit with our expectations or with the conventional wisdom regarding signature whistles. We expected each dolphin to have a signature and for most of his whistles to be the signature type.

But, as we looked at whistles from more recent years, one whistle, which we dubbed the "upcurl," became more and more common for Snubnose, Bibi, and Sicklefin alike. Not every upcurl looked the same. There was still a lot of slop, but they were distinctly similar. In fact, the upcurl gradually became the most popular whistle for all three of the males. If the most common whistle a dolphin produced was its signature, then Snubnose, Bibi, and Sicklefin had all converged on the upcurl as their shared signature whistle.

We went back to our notes and tried to understand why this might happen. In the first few years that we recorded them, only Snubnose and Bibi visited Monkey Mia. They spent some time together, even offshore, away from the feeding area, but they also spent quite a bit of time apart, with various other dolphins. As time went by, they spent more and more time together, becoming an established alliance. We saw them herding females together for the first time during the second year we recorded them. By our third recording year, Sicklefin had joined the alliance and the threesome was herding females almost every day.

The development of a common signature had paralleled the for-

mation and maturation of their alliance. The more they spent time together and cooperated with each other to herd females and to fend off
rival male alliances, the more they had converged on the upcurl whistle.
Was this an alliance signature whistle? It might make sense for male
dolphins to identify themselves as an alliance rather than as individuals
if by doing so they made a statement to each other about their commitment to the alliance ("I am no longer I—now we are we"). A "solidarity"
whistle could inform other males ("Hey, we are very together, so don't
mess!") or it could inform females ("We are incredibly good at being an
alliance. If you mate with us, your sons will have the right stuff"). All
possibilities, and not mutually exclusive ones at that.

We had stumbled across a natural situation in which the unusual
dolphin capacity to learn and modify whistles, first discovered among
captive dolphins, was being put to use in the wild. And not too surprisingly, it was in exactly the same context where the dolphins exhibited
some of their most phenomenal social behavior—namely forming complex and cooperative alliances.

Even now, after spending more than fifteen years listening to dolphins, I am still amazed by the diversity of sounds they produce. I
keep hearing new sounds, no doubt partly because my ear is becoming
more refined, but also, I suspect, because the dolphins keep coming up
with new ones.

I was not the first to be so impressed. Early researchers like John
Lilly speculated that dolphins might have a language akin to our own.
After all, they are smart and social animals. If they make so many different sounds, and we assume that each sound has a different meaning,
then they must be engaging in some very complex communication.

Although at first it seems straightforward enough to ask whether
dolphins have a language, a moment's reflection reveals otherwise. We
use the term *language* first and foremost to refer to what people do,

which is really quite miraculous, although we tend to take it for granted. In his book *The Language Instinct* Steve Pinker put that into perspective for me when he wrote: "Simply by making noises with our mouths, we can reliably cause precise new combinations of ideas to arise in each other's minds." (page 15)

We achieve this miracle by using discrete units (words), each of which has a totally arbitrary, conventional relationship to its meaning. "The word dog does not look like a dog, walk like a dog or woof like a dog, but it means dog just the same. It does so because every English speaker has undergone an identical act of rote learning in childhood that links the sound to the meaning." (Pinker, page 84)

The tremendous potential of language comes about not from having many arbitrary names for things, but from the ways in which we can combine and recombine those words in an infinite variety of ways such that each different order has a different meaning. We can derive meanings out of all those different word combinations because we have a generative grammar: a set of rules about how meaning changes with different word orderings. We know that "Dolphin bites man" is not the same as "Man bites dolphin." Although many people describe "thinking in pictures," we are such linguistic beings that for the most part, we "think in words." We even have special parts of the human brain, Broca's and Wernicke's areas, that are devoted to language processing.

The study of animal communication is still in its infancy, but so far we have not found anything like the infinite communicative potential that human language encompasses in the brains or behavior of other animals, not even among our closest primate relatives. The closest we have come is to find languagelike features in the communication systems of some other species. For example, in their studies of the alarm calls of East African vervet monkeys, Robert Seyfarth and Dorothy Cheney found that the monkeys had very different alarm calls for different types of predators: if an eagle flew over, the monkeys would make a low-pitched, staccato grunt; if a leopard was prowling nearby, a short tonal call; if a snake was slithering through the grass, a "chutter." When they played back recordings of different types of alarm calls, the monkeys would react in a manner appropriate to the type of predator the call referred to. If the researchers played back an eagle alarm, the monkeys would look up at the sky and crouch down low to the ground; a leopard alarm, and the monkeys would run up into the trees; a snake alarm call, and the monkeys would stand up on their hind legs and look around on the ground. The monkeys showed every indication that the calls had different meanings, and, as with words, the relationship between the sound and its meaning was simply an arbitrary convention that young vervets had to learn. But they still come nowhere near having almost infinite potential based on a generative grammar.

Studies of brain activity by vocalizing animals have only confirmed the differences between human and other communication systems. For the most part, vocalizations of nonhuman primates and other creatures seem to be under the influence of subcortical structures, like the limbic system. This suggests they are involuntary responses to emotional states, not the cortically controlled, consciously well-reasoned, carefully articulated, logical utterances that we all know to be typical of human communication.

At this point, it is impossible to get any realistic grasp of how exceptional dolphins may be in this respect. Lou Herman and his co-workers in Hawaii have taken one interesting approach to this issue. They've trained two dolphins, Phoenix and Akeakamai, to use an artificial language (much as some researchers have done with chimpanzees). The dolphins have shown that they can learn to associate arbitrary sounds (in the case of Phoenix) or gestures (in the case of Akeakamai) with objects in their tank (hoop, ring, ball, surfboard, person, and so on). They also have been trained to make sense out of different word orderings ("Bring the ball to the surfboard" means something different from "Bring the surfboard to the ball"). This is certainly impressive and demonstrates something about the cognitive potential of dolphins given proper training; but it is one thing to train an animal to respond to commands from a human using an artificial language created by humans and quite another for an animal to spontaneously generate and use its own.

Language or not, there is no doubt in my mind that dolphins are extraordinarily expressive. The first time I listened to the communication sounds of a group of socializing wild dolphins, I was transfixed. I had listened once or twice to dolphins at the beach at Monkey Mia while they were being fed by tourists and had heard a few whistles and echolocation clicks, thrilling enough at the time. But the racket emanating from this socializing group was both a shock and an inspiration. There were whistles and clicks, and screams and screeches, squeaks and squeals, barks and blats, chirps and "chimp squeaks," growls, grunts, and grates, "balloon rubs," blorps and bangs, heehaws, hurrahs, cracks, and pops. There were sounds I cannot describe or name and seemingly infinite variations on each and every one. All of theses sounds other than whistles are called "burst-pulse sounds" (because they comprise bursts of pulses, pulse being another term for clicks). Almost nothing is known about that whole huge expanse of the dolphin vocal repertoire.

They must be making all those different sounds for some reason. If each conveys some different message, then these animals must be conveying a *lot* of messages. But figuring out just what they are communicating about and how they are doing it is profoundly difficult. Imagine you are an alien from another planet trying to make sense of a human conversation. There you are, sitting in someone's kitchen while he or she converses with a friend, and you listen. A stream of sound emanates from one person, followed by a stream of sound from the other, and so on, back and forth. You have no idea what counts as a "word" or other meaningful unit. With luck and perseverance, you might eventually start to recognize some sounds that you have heard before. You probably notice that the communicators are taking turns for the most part. You might try to learn something about the meaning of the exchange by watching the actions of the people as they converse. Say one person makes a pot of coffee and offers a cup to the other, does that mean they are talking about coffee? After pouring the coffee, they both sit at a table, facing each other, and wave their hands around a bit and move their faces into different expressions. They could be discussing the intimate details of their sex lives or the state of government leadership and you wouldn't have a clue based on their concurrent actions. Where would you even begin? Being from another planet, you would have a hard time even guessing at what might be relevant for humans to communicate about.

Studying dolphin communication in the wild is an even greater challenge. Add to the above scenario the fact that it is almost impossible to tell who is saying what, since sounds are produced internally. Then add yet another problem. Dolphins have attuned their hearing to sounds at frequencies many times higher than anything we can hear. Young humans with good ears can barely make out a high-pitched tone at about 20 kilohertz. For dolphins, that's only the beginning: high-pitched to a dolphin is more like 150–200 kilohertz; comfortable is about 60 kilohertz. There is little demand in the world for tape recorders that record sounds beyond the range of human hearing. A few such tape recorders have been developed for special applications, but they are extremely delicate and expensive machines that don't take well to saltwater, sand, or bouncing around on a boat. Without such specialized equipment, it is impossible to get an accurate sense of the full spectrum of dolphin sounds.

It is no wonder that we know almost nothing about dolphin communication. While most dolphin whistles are at the lower end of the frequency spectrum and can be reasonably recorded with normal equipment, the rest of the dolphin vocal repertoire is at least partially out of range and therefore difficult to properly record and analyze.

Having spent thousands of hours listening to dolphins, I believe they are not engaging in highly intellectual discussions, full of reference, metaphor, and subtle innuendo when they use burst-pulse sounds. But I can easily attribute emotional states to the sounds I hear. Growls sound angry, screams sound freaked out, squeaks sound friendly, submissive, or insecure, grunts sound surprised or threatening, and blats sound indifferent. Burst-pulse sounds seem to convey something akin to pure raw emotion.

This point was made very clearly one day when Snubnose had what was unmistakably a temper tantrum. He had been exceptionally irritable all morning, snapping at people who tried to touch him, pushing and jostling for position among the other dolphins in his attempts to get to the fish that people were offering, thrashing his tail around, refusing invitations to play with seagrass. Then an innocent tourist waded into the water with a bucket of fish. Nicky was right next to this person, but Snubnose barged in, stranding himself in the shallows in his attempt to get in between the tourist and Nicky, but the fish went to Nicky anyway. Snubnose tossed his head. Another fish went to Nicky, and Snubnose tried to roll toward the fish held out at arm's reach. But Nicky got the next one, too. That was it. Snubnose launched into a tirade of screeching, screaming, growling, and groaning such as I had never heard before. He rolled onto his side and then onto his back, with his pectoral fins jutting stiffly into the air, just like a child throwing a temper tantrum. *Grrrrrrrrrr, grrrrrrr, grrrrrr, grrrrrr, hehaaaa, hehaaaa, hehaaaa, bvvvvvvvvvv, bvvvvvvvvv, bvvvvvvvvv, umf, umf, umf, umf, umf,* and so on he went. A few moments later Snubnose lunged after Nicky, jawing at her and chasing her offshore. I don't think he was engaging in anything like a discussion of his feelings and thoughts regarding the distribution of fish and the behavior of tourists and Nicky's appetite, but he certainly made his feelings known.

Burst-pulse sounds obviously contain information about a dolphin's emotional state. It is easy to read into these sounds various states of mind, like anger and irritation, affection and interest. We can't be sure whether the emotional states we attribute to these different sounds are correct or not, but chances are that we do in fact share many of the same emotions with dolphins and that we may have roughly similar ways of expressing our feelings. We grunt with disgust, moan in distress, raise the volume and pitch of our voice as we get nervous and excited, lower the pitch of our voice in anger. Communicative and expressive, but still not language in human terms.

If dolphins have anything truly analogous to language, I suspect it will be manifest in their whistles more than in burst-pulse sounds. If a rise-fall-rise whistle means "Puck," a rise-flat-fall whistle could certainly mean "shark" and a rise-fall-flat mean "pink snapper." In other words, if dolphins use different whistle contours to refer to each other, it seems to follow that they could also use certain contours to name things in their environment. Certainly there is room for such a vocabulary within the huge variety of different whistle contours we recorded. And if they do have "words" for things, the big question is whether or not they have some sort of grammar that enables them to derive different meanings from different whistle orders.

Dolphins vary their whistles in certain ways, almost as though there are rules that would give the variations meaning. The contour might be shortened or lengthened in time, raised or lowered in frequency. Parts of the contour are sometimes deleted or repeated or extra bits are tagged on. If all of these modifications have particular meaning to the dolphins, it seems to me that the sort of infinite potential through recombination characteristic of human language could to some extent exist in the communication system of dolphins. It will take far more than this kind of speculation to truly demonstrate language in dolphins, fun though it is to speculate. What would dolphins talk about if they could talk to us? Would they discuss the latest good fishing spots or tell stories about their encounters with sharks? What if we could ask them questions like: Are you really as smart as we think you might be? Can you stun fish with your clicks? What do you do with those sponges? What could they tell us about themselves and about their ocean world?

I had a vivid dream once in which I was able to converse with a dolphin. I was in a bar, a seedy, smoky, out-of-the-way joint of the sort that serious alcoholics tend to frequent. I was sitting at the bar enjoying my double-malted Scotch on the rocks, feeling a little uneasy, a woman alone among a lot of drinking men.

Some guy seated at the bar next to me struck up a conversation, I don't remember about what. I looked away from him for a moment, and when I looked back, he had turned into a dolphin! He still sat at the bar stool, coolly sipping his drink through his long beak with his tail flukes resting on the bottom rung of the barstool. I was flabbergasted. My God, I thought, this is *it*. I can talk to this dolphin/guy. I can ask him anything I want to know about dolphins and he can just *tell* me the answers.

Somehow I knew that I would have only one opportunity and that I had to come up with *the* question that would shed the most light on everything I wanted to know about dolphins. But when I tried to think of what that burning question might be, I drew a total blank. I couldn't even muster a simple inane question, much less one that would be profoundly revealing of dolphin nature.

If you've ever had a dream where you are trying to run away from some monster and your body just refuses to move, then you know something of the sensation I felt. I was terrifically frustrated, but that only made it worse, and then it was too late. He turned back into a human being, and my opportunity was lost.

Although Snubnose, Bibi, and Sicklefin had certainly thrown a wrench into the works by not conforming to the expectations of the signature hypothesis, we did continue to learn the signatures of individual dolphins in Shark Bay. It even became possible to some extent for me to recognize certain dolphins just by listening. I could drop a hydrophone in the water and hear Nicky whistling nearby. Sure enough, she would show up a moment later. Or I would hear Cookie's plaintive whistle off in the distance and envision him heading home to Crookedfin. One windy afternoon when we had opted not to battle the waves, I took the luxury of sitting on board *Nortrek*, peacefully strumming my guitar. Faintly at first, just teetering on the fringe of my awareness, I heard a whistle. It grew louder and louder until I recognized Smoky's signature. I could hear his whistle being transmitted right through *Nortrek*'s hull. I jumped up and went out on the deck. Sure enough, there was Smoky. He circled the boat, chasing little fish belly up, then he whistled once or twice more and headed out to the northeast, no doubt en route to Yogi.

Learning to recognize some of the dolphins by whistle gave me the feeling that I was privy to a whole new dimension of their lives. I could hear who was around, just as they must do. They undoubtedly hear who is where and probably make some good guesses about what they are up to as well. Up to that point, in my visual imagination, the world of the dolphins had in some sense been confined and claustrophobic. I had imagined them virtually blind in the opaque water, holding their breath, able to echolocate only within a narrow window straight ahead, always searching for each other in the dark. Not at all. In fact they are a community, bound together through the world of sound.

I and other researchers have put many, often tedious hours into analyzing dolphin communication, always with the hope of "breaking the code" or making some key discovery that will allow us finally to pull back the curtains and understand these animals. But I have also wondered what the dolphins think of *our* communication. They hear us speaking constantly. Do they recognize that there is something special about that? The closest I have come to an answer to that question came one day when I was swimming with Nicky. It was always a challenge to engage her attention in a good swim. I suppose she knew I wasn't likely to have a fish handout, and I had the impression that she found the flailing swimming motions of humans irritating. While Puck was usually game and would often approach, whistling, and swim in circles around me or allow me to put a hand against her side, Nicky almost invariably ignored me. But on this occasion I was able to really capture her attention. I got into the water and she gave me an "Oh, it's just you" sort of look and headed away. I searched my brain for some inspiration, something really different and interesting that I could do to interest her. On the bottom just below me I noticed a stick, about a foot long and perfectly straight. I dove down and picked it up, not sure what I would do with it. The feel of it in my hand gave me an idea. I started to write with it on the sandy bottom "I LOVE NICKY" in big stick letters.

Nicky turned, circled me at a bit of a distance, and then moved closer. She came right alongside me and put her face up to the tip of the stick, her head moving in synchrony with the motion of my hand. Her attention was riveted as she followed every movement intently. I wrote my message to her, moved over to a fresh patch of sand, and wrote it again and then once more. She followed me each time. When I stopped several minutes later, her eye rolled to meet mine, then back to the writing in the sand; she paused and only then swam away. I had never before, and have never since, held her attention so completely.

I was dumbfounded—not just because I had won Nicky's attention, but because she seemed to realize that this behavior of mine, writing, was something significant. Something worthy of her attention. I don't begin to imagine she understood the meaning of my message to her or that she truly understands how profoundly important writing is to the lives and cultures and communication of humans. But her extraordinary attentiveness to my stick writing made me think that she recognized at least that something truly important was going on. That is exactly how I feel about the communication of dolphins and why I chose this difficult topic as the focus of my research; it is difficult if not impossible to understand, but there is definitely something profoundly interesting going on.

LIFE AT MONKEY MIA

When we had arrived at Monkey Mia in 1990 to begin our study of mother-infant whistle use, we found the camp basically shut down and a herd of bulldozers noisily digging up the place. Monkey Mia was being transformed from a rough little fishing camp to a resort, complete with tennis courts, a swimming pool, a tiki hut serving cappuccino, and a set of chalets to be rented out to visiting tourists. It was a transition that was probably inevitable. Monkey Mia's dolphins had become the top tourist attraction in Western Australia. But it was also a transition we faced with much trepidation. More and more tourists meant more and more development and, inevitably, more and more risk to the dolphins and their bay.

With the new development, I knew that life at Monkey Mia would never be the same. I had never objected to "roughing it" in the field. In fact, life at Monkey Mia was far more convenient and luxurious than life in my VW back in California. In the beginning we had lived in small sleeping tents, cooking over a fire. Later we graduated to larger tents, the sort one could stand up in, with a table and cots. After weathering some severe wind and rain in those big tents, which performed like genoa sails in a strong tailwind and leaked without fail, we had invested in a "caravan" (mobile home trailer) with a canvas annex attached to one side. This was really the lap of luxury, replete with furniture, a propane stove, and solid walls to keep the sand and dust out of our equipment.

I have always felt more comfortable in a wilderness than in a city, so the isolation of Shark Bay was in no way intimidating, and I didn't mind washing in salty water, sleeping in a tent, or washing dishes in a bucket. The one thing I did find intimidating about living at Monkey Mia was the weather, a fear that was deeply instilled in me on May 21, 1988.

I am awakened several times in the early morning hours by the sound of flapping canvas: the wind yanking at the loose door flap of our annex (a canvas tent attached to the side of the caravan). I don't want to wake up even though I am dreamily aware that the wind is strengthening and I need to batten down the hatches and check on the boat. I know I anchored it solidly this afternoon, but still I envision it tossed off its moorings, smashed against the beach, or set adrift in the stormy seas, en route to Madagascar. I really don't want to get up and go out there, but the wind keeps howling and everything is shaking, rattling, and flapping. When a metal cookpot is knocked off the kitchen table, I am fully roused. The campground generator is off, so I fumble for the flashlight, which I keep near my bed. At least it wasn't the pot of leftover spaghetti.

I step outside to find others are already up and moving around the camp, checking on their boats and securing tents. Flashlight beams shift and crisscross up and down the beach, occasionally silhouetting a human figure, a caravan, or a tent. My *God*, the wind is positively blasting out of the northwest. That is where the really big storms come from: straight off the ocean.

Wind and water roar, whipping my hair around my ears and into my eyes. Down at the water's edge, my little dinghy is careening around over the waves, rearing straight up against the leaden sky, and then crashing back down. I hope it's not banging against the bottom.

A gust nearly knocks me over and sends sand stinging against my legs. I hear crashing sounds around the camp: garbage can lids and other loose things blowing around. My most immediate concern is the caravan annex. The canvas won't withstand too much more of this beating before it will rip wide open. One of the front poles has slipped out of place and is no longer providing support. I pull it back up and try to poke the nipple through the grommet hole, but the canvas is flapping so wildly that I cannot maneuver it into place.

The annex is full of tables, dishes, pots and pans, buckets, boat tools, desk, books, boxes of papers, electronic gear, my clothing, and bed. I stand, bracing against the force, leaning into the tent pole, considering what is going to happen when it rips wide open and the contents are exposed to the elements. But I can't remain here trying to hold the pole up, so there's no alternative. I let her rip and begin a mad scramble, trying to shove papers and cameras and anything else I can salvage into the caravan as the windward panel of canvas tears loose and comes under complete control of the wind.

I hear the muffled thrumming of the generator motor turning over, and then the lights flicker on. The generator is usually off and the camp is dark at night. My heart leaps and I feel suddenly queasy, because I know that the Masons wouldn't put the generator on at this hour unless there was something really serious. My fears are confirmed.

A cold rain is pelting down as I run back to check on my boat again. The dinghy is pitching about wildly and already taking on water, making it much heavier. With all that weight, the anchor line, which now seems pathetically inadequate, is likely to break at any moment, in which case the boat could come careening down on top of me. For a few moments I stand shivering in the wind and rain, with tears pushing out of my eyes and blowing off my cheeks, considering what to do. Then Craig, the Masons' son-in-law, shows up. He is usually an extraordinarily calm, easygoing, good-humored chap, but I can see immediately that even he is worried. Yelling above the wind, he offers to help me take the motor off. The dinghy itself is tough aluminum and can withstand a beating. The motor, if it gets submerged, will be ruined. Together we wade in and muscle the motor off, and Craig hauls it up to the only solid building around: the cinder-block toilet house.

On the way back to the caravan, I pass by the camp of the elderly couple whose company I have enjoyed these past two weeks or so. They are in their eighties and have been coming to Monkey Mia occasionally for many years. Their tent is now a flapping heap of soaked canvas, held to the ground by a few remaining stakes, with debris strewn all around, and they are huddled in their car, wide-eyed.

The storm is pushing waves up the beach, and by the time I get back to the caravan, the annex is completely down, pots and pans and dishes strewn around on the ground. A box of sodden books sits by my bed, which is collecting a puddle amid the sheets and blankets. Waves are breaking menacingly just a few feet from the front of the caravan.

Salty flecks of wind-whipped foam are blowing off the tops of the waves. I realize at this point that the only thing left for me to do is get the car, pile as much as I can into it, and then drive back away from the exposed beachfront. I will have to abandon the caravan, the boat, and everything else and hope for the best.

Nicki-the-human shows up to help me carry my equipment to the car: the computer with all our records, notebooks, cameras and binoculars, video equipment, some file folders with important records, and a few books. By then waves are breaking against the caravan, and we are wading in and out with piles of stuff held above our heads.

Nicki and I dive into the station wagon and drive around to the lee side of the toilet house. Here, compared to the beachfront, it is relatively calm. We turn off the engine and sit huddled, wet and anxious, behind the steaming windshield of the car as it shudders in the gusting wind. Debbie pulls up alongside in her little camper-van. We all sit and stare, dumbfounded, as debris comes flying over us—garbage can lids, bits of tents, and clothing. There is an ear-splitting din as things crash and bang and collide.

At dawn the sky is the color of dried blood, and an hour later the din fades and there is a sudden and eerie calm. Tentatively we open the doors of the car and clamber out. Other people too are emerging from their hiding places. The quiet is stunning, the beachfront barely recognizable.

Not a single boat remains in the mooring area. Several lie overturned and washed up against the stone wall, with their motors strewn in bits around them. One boat has blown right up over the stone wall, onto the upper level of the campground. The jetty is gone but for a few broken pilings sticking out of the water and some boards strewn on the beach. There is debris everywhere amid great mounds of seagrass and sand. The water has abruptly receded. It is as if the bay has emptied, a huge backwash after the surge that has left the nearshore flats exposed. Just offshore are the upside-down hulls of several boats, floating amid a tangle of ropes and debris that seems to be holding them all together. I recognize my boat among them. Farther offshore, the water is still churning wildly, spitting up an eerie white mist.

Through all of this, I've had a few fleeting moments of concern for the dolphins, but mostly I've been too distracted. Now I have the chance to worry. How could they possibly catch a breath of air when the water is so wild? Would they get washed up on the beach?

Dolphins generally don't last long out of water. Because they have adapted to the constant cooling effects of water, they overheat without it, and their skin quickly dries out and blisters in the sun. Similarly, because they are accustomed to the reduced effects of gravity in water, the sheer weight of their bodies on land can crush internal organs. Local legend has it that after the last time a cyclone came through Shark Bay, several dolphins were found washed up on the beach by Denham, buried under piles of seagrass that provided moisture and protection from the sun.

I wander down the beach immediately adjacent to Monkey Mia and find everything but dolphins: huge piles of seaweed, pieces of boats and camping gear, dead fish, and the remains of other sea creatures. It is going to take a monumental effort to clean all of this up. But we are lucky: the buildings are still standing, and nobody was hurt.

As we are surveying the damage, the wind picks up again. Now it is coming from the south, blowing out to sea from behind us. It picks up alarmingly fast, and before we know it, all the garbage can lids and pieces of junk that had been blowing past us earlier come flying back in the other direction, heading seaward. Nicki and I run back to the station wagon and jump inside. We had experienced a lull in the storm as the eye of the cyclone passed over, and now we were feeling the other side of that colossal, rotating mass of air.

A couple of hours later we reemerge. The sky is clearing, though the light is somehow eerily subdued. Everyone is out wandering around, looking stunned. A crowd of oldies has gathered, and they are engaged in animated conversation. Already they seem to have come to terms with their losses and even laugh as they tell their stories. They are a resilient bunch for sure, accustomed to the bizarre and terrific forces of nature that Australia offers to its inhabitants. This storm had been heading off the coast somewhere far to the north of Shark Bay when it suddenly veered around and came smashing back toward land directly at us. The weather service had somehow missed it, and therefore we'd had no warning: bad news for us and worse for the fishermen in Denham whose boats are their livelihood.

A remarkable solidarity developed among people at Monkey Mia over the days following the storm. Those with living spaces still intact offer beds to those without, people share food and help each other gather up and sort through the debris. The more transient visitors, those tourists unfortunate enough to have visited Monkey Mia at just the wrong time, have mostly piled themselves into their camper wagons and left Shark Bay, probably never to return.

Nicki-the-human and Debbie and I are among the only young folk around crazy enough to swim in these shark-infested waters, with snorkel and flippers. We become indispensable to the old fishermen, eager to find equipment that has been lost from their boats and must now be lying on the bottom of the bay. We spend hours swimming slowly back and forth, scanning the bottom, and succeed in finding some toolboxes, fishing nets, pieces of outboard motors, and broken mooring ropes. Much of it is ruined by the saltwater, but some is still good. My boat has fared well. After the storm I flipped it back over, untangled the mooring ropes, and put the motor back on. The only loss was a tool kit.

A week or so after things have settled back down again, I wake up early and go down to see the dolphins. Holly is in at the beach and in a languid mood. I have the urge to jump into the water with her this morning, so I don my snorkel and mask. She stays just offshore, watching me get dressed, and I can tell by her patient, attentive waiting that she too is in the mood for a swimming partner. In my excitement, I fall over trying to get my flippers on. Holly is whistling as I slide into the water alongside her. Meeting her in her own element is quite a different experience from the usual Monkey Mia meeting. Rather than awkwardly stranding herself in the shallows, straining her neck to peer up at her human visitors, she is graceful, easy, and in command. She sidles up alongside me, moving very slowly. Her eye, inches from mine, is half-shut. She is so calm that I, in turn, relax. I reach out and place my hand on Holly's side. Her eye opens just a bit, rolls back to look at me, and then returns to its half-closed position. She remains close. I wrap my arm around her, and she accepts even this. Side by side we progress slowly out into deeper water.

She is so relaxed and gentle and warm. I soon match her mood, drifting along in a dreamlike calm. Then she gently moves out from under my arm and heads down toward the bottom. We are in about twenty feet of water, and I try to dive with her, but my awkward flailing seems inappropriate so I retreat to wait at the surface. Below me, she is poking at something on the bottom, but the water is too murky for me to see. A moment later she comes back up toward me, dragging something large, white, and apparently heavy, which she is holding in her jaws. She comes directly to me and delivers a plastic bag into my hands. I take it, and she moves away, diving again at some distance from me. She is done swimming with me for now, and it would be useless to try to catch up to her. Besides, she has put me into such a relaxed mood that I don't feel like exerting myself. I tread water for a moment and untie the plastic bag. It looks vaguely familiar somehow. Inside are a ratchet wrench set, pliers, screwdrivers, some spark plugs, and flares. It is the tool kit from my boat.

The cyclone was, in many ways, a turning point in relationships with the Monkey Mia oldies. Whatever differences and misunderstandings that had developed were largely washed away by the storm. We had all been put in our place by a force much larger than our petty differences. Things would never turn rosy with old Bondi, but we suddenly gained some acceptance with many of the old-timers. It was as though surviving the storm were a rite of passage into the community.

These old folks were accustomed to living in harsh and isolated conditions, and they looked after each other. From them we had learned how to do some basic repairs on our boats, catch fish, splice a rope, and tie a few useful knots and what to watch for in the weather. Now they also kept our dinner table supplied with fresh snapper on a regular basis and provided good company over a beer or two on many evenings. Part of the ritual of evening life in those days was to wander along the beach, stopping at each of the cleaning tables to admire people's catches and talk about fish and the weather as pounds of fat pink and black snapper, bluebone, mullet, and whiting were scaled and gutted for the freezers. I felt a certain tension in my shoulders release and dissipate as we finally became part of this tightly interdependent, albeit remarkably quirky, community.

Thus it was a shock when the Masons moved on and the camp changed hands and was turned into the "resort." We found ourselves surrounded by people we didn't know and with whom we felt no history and little connection. Clean, businesslike young faces dressed in sparkling white T-shirts and creased shorts replaced the tattered, gray-haired, unshaven, fishy-smelling, and barefoot characters we had come to know and appreciate. In subsequent years, once in a while, I would encounter one of the oldies camped somewhere on the grounds of the resort, looking very out of context. One by one they stopped coming altogether. Some got too old to make the long journey, and others complained that the resort prices were beyond their means. In any case, there was no new generation of dedicated fisherfolk to replace them, and the face and character of Monkey Mia was forever changed.

A few, people who worked for the resort, stayed for months or even years at a time, creating some sense of community. But mostly an ongoing flow of humanity from all over Australia and the world came to see the dolphins, passing through, making use of the facilities, and then heading off down the road, never to return again.

As more and more people came to visit the dolphins, the interaction became, by necessity, more and more closely managed. A team of rangers had been hired to ensure that nobody, dolphin or human, got hurt and to answer visitors' questions. This was critical. Things had gotten chaotic as large crowds of tourists scrambled to touch and feed the dolphins.

Watching the interactions of people and dolphins was an ongoing source of both insight and entertainment for us, ranging from the sublime to the ridiculous to the downright dangerous. For some people, interacting with a dolphin is a highly emotionally charged affair. I know some of my own thoughts and feelings about interacting with the dolphins went way beyond what would transpire in an interaction with the neighbor's dog. I remember some of my feelings during one of my first encounters with Nicky. I was standing on shore, and Nicky was floating just five feet from me, on her side. One eye was tilted up and watching me. I waded into the water. "Nicky! Good morning, gorgeous. Are you waiting for something?" Her eye followed my hand, then rolled up to look me in the face, back to my hand. I reached toward her to stroke her side, and her jaw opened just a little bit, then stopped in a slightly open position, revealing her perfect row of teeth. She must have thought I had a fish offering, then realized it was just my hand. Her body drifted away from me a bit, and I took a step closer. She was watching someone else walking on the beach. I could imagine her thinking, Does that one have any fish? This one is clearly a loser. Just wants to paw at me with those dangly mitts.

Her tail flukes bumped my shins as she moved off to approach the newcomer, a gangly boy of about fifteen. The kid was tentative. He looked at her but stayed back, holding his arms stiffly at his sides. His demeanor revealed a combination of amazement and fear. Nicky rolled on her side and watched him. We all stayed put and waited. Holeyfin surfaced about fifty feet off, foraging. She took a couple of breaths, and then her tail came out as she dove. Nicky suddenly turned out toward her, as if she had heard something interesting, but then stayed with us.

It was a peaceful scene on the beach, just two humans and a dolphin, quietly enjoying the moment. But I was quick to become restless. I wanted Nicky's attention. I wanted to play with her, to touch her, to explore her responses. And if she responded badly to me, I would be embarrassed and mortified. I wondered if the kid felt the same way. Maybe part of his fear had to do not so much with concern about getting hurt as with whether or not Nicky would like him. What could be worse than being disliked by the most altruistic and loving, perpetually smiling, and happy creature of all? I approached Nicky again, and she swirled her whole body around in order to see me. I spoke to her, saying ridiculous things intended to convey my friendly intentions. She stayed put as I came close and paused, then reached out to stroke her. She suddenly jerked her head out of the water toward me, clearly agitated. I was heartbroken. I backed up out of the water, chastened, and stood still, telling myself that if I had a fish, she would have been nicer to me. Maybe she just didn't like the color of my purple shirt.

The kid decided to have a go. He approached Nicky, and she moved closer and poked her rostrum against his knees, rolled her eye up at him, and emitted that questioning little *eheh?* vocalization that the dolphins always make when they are asking for fish. There was no sign of the irritability she had shown toward me a moment ago. She allowed him to stroke her side, leaning against his thighs, then nibbled at his toes and moved away. I was intensely jealous.

For many visitors, perhaps even the majority, the goal is to get a photograph taken of themselves feeding a fish to a dolphin. Wading in tentatively, fish in hand, they dangle the fish just in front and out of reach of the dolphin while a partner focuses the camera. Once the photo has been taken, it's time to hit the road. "Been there, done that, checked it off the list." In recent years, tour bus operators have begun to carry loads of people up from Perth. After the long trip, the passengers are discharged onto the beach for an hour or two, plenty of time for that photo opportunity and a visit to the snack bar. Then they load up for the ten-hour journey back to Perth.

I feel sorry for these folks. They have made such a long trip out to Shark Bay, and their experience seems so impoverished. I imagine they must feel gypped, but perhaps it is all they expected. Once I watched a chap pull up into the parking lot in his land cruiser, roll down the window and assess the situation, then drive directly down onto the beach in front of the dolphins, snap a picture through his car window, and drive away.

What disturbs me most about these drop-in tourists is that they don't seem to realize just what an incredible privilege they are being granted. Through their eyes the dolphins are reduced to quaint little critters, put here to provide a trifling amusement and a photo op.

Fortunately rare, but much more deeply disturbing, are the really ugly tourists. Some of the more benign don't think twice about throwing their trash around on the beach and even in the water among the dolphins. Or they chase after the dolphins, yelling and grabbing at them, generally disrupting the whole affair for everyone. The worst of them can be downright dangerous.

One day in 1986 (before the dolphin-human interaction came under more thorough control of the rangers), as I was standing among the dolphins, recording with my hydrophone, a wiry-looking man with very short hair and a loud voice came down the beach. He was with a woman with enormous hair and a very revealing neon pink bathing suit. The loud man strode into the water and unceremoniously grabbed hold of Holly (she was three at the time). He picked her right up out of the water under one arm as she writhed and struggled. But the guy was big and completely oblivious of everything but the camera in his girlfriend's hands. Impatiently and at top volume, he demanded that she get the camera focused and take a picture. The girlfriend struggled with the camera, whining and fumbling, while Holly, eyes widened in terror, began to slip from his grip. He hung on, maintaining his pose with a facsimile of a smile pasted to his face. By this time I was pounding him on the back with both fists, but he was completely unaware even of that. I suppose when they got their pictures developed they probably wondered about the angry-looking woman with clenched fists standing directly behind him, making a mess of their photo.

Photo taken, he dropped Holly abruptly into the shallow water, then grabbed her dorsal fin and gave it a jerk, a gesture akin to a macho "slap on the back." Holly fled to Holeyfin's side, whistling intensely. When I gave this guy a piece of my mind, he seemed surprised, and afterward I heard him complaining to his girlfriend about the "damn Yank who thinks she can tell me what I can and cannot do . . . oughta go back home to America." Oh, well.

On the other hand, there are many visitors for whom seeing, touching, and feeding a wild dolphin is the fulfillment of a lifelong dream. One incident among many stands out in my memory: An old couple stands with just their toes in the water. The woman's eyes have the pale and disarming look of a blind person, and she is holding her husband's arm. Her face is positively brilliant with excitement. Her husband very gently helps her into the water, instructing her to pull up her pant legs. She is so excited that she can hardly stand to be bothered and starts to lose her balance a bit. He steadies her with his shoulder and leads her forward toward the dolphin. She is reaching out toward the general direction of Puck, fingers pumping, seeking. Her face is quivering with excitement; her eyebrows, raised high with elation and expectation, smooth some of the many wrinkles on her aged face. The emotion is spelled out on her face, uncensored in the way of the blind. Hand groping, she says to her husband, "Oh, is there really a dolphin? I want to touch it. Lead me close so I can touch it."

Everyone else stands back in deference to this couple, clearing the way so they can approach Puck. The blind woman is still leaning against her husband, who is completely and gently attentive to her. She leans forward and reaches out. Her husband takes her hand and carefully guides it to Puck's side. The woman's face registers a thousand feelings in that instant. She is surprised by the warmth. Her hand, filling in for her lack of vision, explores Puck's form, and Puck, seemingly aware that this is a special occasion, tolerates the groping. The husband talks to his blind old wife, holds her arm, directs her hand, then after a time leads her gently back to shore. Her pants are soaking wet. There are tears rolling down my face.

The Monkey Mia dolphins are experts in human behavior. They have seen it all, but they still take an interest in anything out of the ordinary. I have seen them examine at great length pregnant women, babies, people on crutches or in wheelchairs, or people who just plain look weird.

One day I watched Nicky exploring the wonders of human

anatomy. It was after a feeding session on a calm day with a small group of easygoing tourists. Nicky and Puck were swimming slowly back and forth among the people, some of whom had drifted out into waist-deep water to cool down. Puck had approached a middle-aged woman and was "parked" in front of her, mouth agape, face out, lingering there and allowing the woman to stroke her side. Meanwhile Nicky had circled behind the woman, who was bending over toward Puck. Nicky stopped right behind her, staring intently at the woman's bottom, turning her head to look first with one eye and then the other. It was, in fact, an amazing bottom: tremendous, with lumps and rolls and bulges over-flowing from a too tight bathing suit. Nicky continued her examination, putting her face up to within a few inches of the woman's bottom and growing wide-eyed as the woman moved slightly, reaching farther toward Puck. The woman was unaware of the attention she was receiv-ing from behind, even when Nicky reached out her pectoral fin and stroked the woman's bottom several times. Nicky decided to investigate further. She moved even closer, poking her rostrum into one bulging cheek and then backing off to inspect once again. I could imagine her thinking, Wow, look at this! What does it feel like? The woman re-mained oblivious through it all.

Perhaps because of their preconceived notion that dolphins are per-petually sweet and kind and gentle, people are sometimes remark-ably unaware of dolphins' signs and signals of irritation. When a dolphin doesn't want to be touched, it tosses its head, jerks away, moves off, or snaps at the offending hand. These signals are really much the same as one would expect from a dog. But somehow people tend to ig-nore them and to persist in their attempts to touch the dolphins. What they don't realize is that the dolphins are quite capable of biting, hitting, and knocking people around if they get a mind to do so.

At one stage, while the three adult males, Snubnose, Bibi, and

Sicklefin, were frequenting the Monkey Mia shallows (males tend to be more aggressive), bites, sometimes serious enough to require stitches, were practically a daily occurrence. What provoked these bites varied tremendously.

Larry Richards, Andrew's dad, came to visit Monkey Mia one year. He walked into the water and ended up standing next to a tourist who was aggressively and persistently rubbing Holeyfin's side, using his fingernails as if scratching a pet dog. Holeyfin was trying to attend to the possibility of a fish handout but was obviously irritated by this rough treatment. She tossed her head and even snapped at the hand a couple of times. The whites of her eyes were showing, but she was trying hard to get a fish handout and obviously didn't want to be distracted. The guy completely ignored her signals and persisted in clawing at her sides. Once the fish were gone, Holeyfin backed up and chomped on the nearest hand, which by then happened to be Larry's. He gushed blood from several deep punctures in his hand. Larry, a zoologist with many years of experience working with wild animals, took it all in good stride.

Some people become peculiarly embarrassed when they are chastened by a dolphin. One man who was bitten by Bibi tried to hide his injury from view, but the look on his face and the blood dripping down his arm were dead giveaways.

I have a few dolphin injury scars myself. One time I was playing give-and-take with a piece of seagrass with Bibi. He seemed to be in a friendly mood, but, with a shifty look in his eye, he dropped the seagrass closer and closer to his face until, when I reached for it, he lashed out and grazed my middle finger with one of his sharp teeth.

On another occasion I was swimming in the shallows at Monkey Mia. There were several dolphins around, all ignoring me as I tried to observe them. Then Sicklefin approached. I felt nervous, as he was notoriously rough with people, and I made no attempt to touch him or reach out toward him. He dove underneath me and very quickly, very powerfully, hunched his back up, causing the blunt, hard front edge of his dorsal fin to punch into my thigh. It was excruciatingly painful. My entire thigh seized up and turned purple, and I still have a dent in the muscle from that injury.

I have to admit that there have been occasions when I felt people deserved to get bitten. One time a particularly large, unruly, and clamorous crowd had gathered in the shallows. Nicky, Puck, Holeyfin, and the three males, Snubnose, Bibi, and Sicklefin, were all around, and all had been showing signs of irritation. Even Puck, usually by far the most tolerant, had swatted a kid with her tail flukes. One woman kept lunging toward Nicky, trying desperately to touch her. Rolls of white blubber shook along her torso under huge breasts that heaved and bulged out of her bathing suit top. She kept shrieking at her two children to get closer to the dolphin for a photo. Each time Nicky came near, she maniacally dangled her hand in Nicky's face in an attempt to get her attention long enough to focus the camera. Finally Nicky, having reached the end of her rope, lunged out of the water, grabbed hold of a roll of midriff bulge, and began shaking the woman like a dog worrying a rabbit. The woman, with a look of terror, fell backward into the water, camera and all.

Then there are the visitors who come with their heads crammed full of "New Age" wisdom about "dolphin consciousness." They come to channel dolphin spirituality, to meditate with them, to achieve enlightenment in their presence, and to learn from them about how to live in harmony with the universe. I appreciate that these people at least feel that the dolphins are something special and are willing to take the time and effort to learn to recognize them and get to know them a bit. But as they work to fit every experience, every action, on the part of their fellow humans and dolphins into some mysterious master plan, it becomes clear that it all revolves suspiciously around themselves and not the dolphins.

Still, some of the New Age dolphin groupies have provided particularly good comic relief over the years. One woman considered herself to have special abilities to communicate telepathically with dolphins. She came to Monkey Mia for several weeks during the early years of our time and spent hours meditating cross-legged on the beach. Sometimes dolphins came, sometimes they left. She seemed convinced that they were attracted to her "mental energy." It looked to me as though they were doing what they always did, but who was I to burst her bubble? One day when she arrived at the beach just a minute or two after the dolphins had left, she threw quite a temper tantrum, stamping her feet and complaining, "And I am supposed to be so psychic."

Then there was the wealthy divorcée who made a living massaging horses. She had an entourage of young followers who, dressed in flowing robes and beads, chanted on the beach, meditated to the dolphins, sipped herbal teas, and munched bean sprouts. She made her first visit to Monkey Mia around the same time as ours, then returned two years later with a group of about eight women. She had swung a deal whereby she had her way paid by "leading" a tour to Monkey Mia. The women all stood in the water in a circle, holding hands and chanting. Old Holeyfin came along and began checking each one out for the possibility of a handout. She swam up to one woman after the other, opening her jaw in the typical begging gesture. The women were thrilled. Holeyfin had "tuned in to their energy" and "was trying to communicate with them" (I don't think they meant "Got any fish?" either). They all had expressions of exultation on their faces and were absolutely a-twitter. I guess they felt that they had got their money's worth.

One woman, attracted to the idea of a "water birthing" and convinced that the dolphins would serve as midwives, arrived at Monkey

Mia during her final stage of pregnancy. Luckily for all, when the local nurse in Denham got wind of this plan, she put her very down-to-earth foot down firmly. She didn't want to have to manage the consequences of such an endeavor and knew full well that the dolphins were unlikely to do so. The water can be chilly and is full of bacteria, not to mention sharks, which are of course attracted to blood.

Over the years, though, a handful of nonscientists have taken great interest in the dolphins and dedicated themselves to establishing a real rapport with them. Spending hours with the dolphins on a daily basis, these people have become astute observers of dolphin behavior, advocates for their well-being, rangers, research assistants, and educators. The rewards for such dedication are in some ways tremendous, in others sparse. The thrill of playing with a wild baby dolphin, of getting a special greeting, of seeing them do something remarkable, of piecing together a richer and more complete understanding of their lives, is hard to beat. At the same time, the dolphins can be exasperating. I've often wanted them to return the same enthusiastic affection I felt for them and instead got barely a glimmer of recognition.

Nicki-the-human developed a special rapport with Puck. Whenever she had the chance, Nicki would jump into the water with Puck, who clearly recognized Nicki and seemed to particularly enjoy her company. The two would dive and roll and touch and move along together. Puck accommodated her swimming speed and style to remain with Nicki. The two even developed their own communications, exchanging sounds and gestures and a style of interaction that was unique and personal.

Denise Myers, a student at the University of Michigan who spent three summers with us working as a research assistant, focused her attention on Sicklefin, the large and often aggressive male whom most of us preferred to avoid. Denise, a buxom, bikini-clad blonde, spent long hours just standing with Sicklefin, sometimes with a hand against his side or with her arms wrapped gently around him, inviting him to play with bits of seaweed, talking to him, and swimming with him. He clearly approved of her attentions. When Denise would wade into the water, Sicklefin would break off whatever he was doing to approach her in an enthusiastic greeting. He tolerated from her things that nobody else would dare even to attempt.

Even a dog, Ringer, became a dolphin groupie. Ringer, an Australian shepherd, fur bleached pale from salt and sun, was never much interested in people, but he spent most of his waking hours standing among the dolphins in the shallow waters at Monkey Mia. Oddly, Ringer never did much of anything with the dolphins; he just seemed to derive intense pleasure from being close to them. He would push up as close as he could get and stand there, transfixed, shivering with tension and excitement and cold. His ears would rotate around when the dolphins vocalized, and I often wondered what he could hear. When the dolphins were not present at Monkey Mia, Ringer would scan the horizon, waiting. The moment he caught wind of them, almost always before any human would notice, Ringer would charge down to the water and wade in, beside himself with anticipation.

One day I found myself alone on the beach with Ringer when he spotted Holeyfin coming. I sat still and watched them. Ringer approached the water's edge and sat down, poised and attentive. Holeyfin approached the water's edge from the other direction and stopped, facing Ringer, similarly poised and attentive. Ringer, in slow motion, like a cat stalking a bird, raised up off his haunches and took one very slow step toward Holeyfin, ears alert, head down and forward, an expression of intense concentration on his face. Holeyfin lifted her head out of the water, looking right at Ringer. She began to emit a litany of dolphin vocalizations, squeaky, splatty, gurgly, buzzy sounds. Ringer's ears swiveled and flicked and twitched. He took two more slow-motion steps toward Holeyfin and stopped again, eye to eye with her, just a foot from her face. I wish I could have read their minds at that moment. What was Holeyfin trying to communicate to this dog, and what did he make of it? The spell was broken when a group of kids came rushing up, splashing into the water and yelling excitedly. Ringer looked sheepishly at the kids and resumed his usual stance, close to Holeyfin, in water up to his chest. His fur was bleached pale from the salt and sun. For whatever his doggie reasons, Ringer was obsessed with dolphins.

When it was decided that no dogs, not even Ringer, could be allowed on the beach, to discourage the tourists from permitting their dogs to run free and chase dolphins, poor Ringer was no longer allowed to visit with the dolphins. He had to be chained to his owner's caravan. After suffering with that arrangement for a while, his owner mercifully sent him off to live somewhere else, far from the temptation of the dolphins. I suspect, wherever he is, he still dreams of them.

While resort life is not really my style, and I miss the old days when interactions with the dolphins were less structured, in retrospect the changes at Monkey Mia—the people and the place—have been not all bad. Most of the horrors we imagined would accompany the increasing development and tourism have not come to pass. The rangers have done an excellent job of looking after the welfare of the dolphins and educating visitors. And in spite of the crowds, I do not begrudge anyone the opportunity to visit with the Monkey Mia dolphins.

Besides the change from camp to resort, our research team also grew over the years. We had made tremendous progress toward understanding the Shark Bay dolphins: we had learned to identify over four hundred individuals, and we knew the sexes and approximate ages of about eighty to one hundred. We had begun to chart their social structure, to understand their sexual habits, to document their hunting and fishing techniques—from sponge carrying to bony banging. We had listened in on their communication and made some important headway toward understanding. With so much background information available, what we had done was to open the door for further studies. There was much, much more to do, and we needed to recruit other researchers with time, energy, and expertise to devote to the dolphins.

But as more people joined our team, each with at least one assistant, living quarters quickly became overcrowded. After spending most of the day on an eight-foot boat with at least one or two other people, I craved some solitude and quiet back onshore and time to take care of desk work: journal entries and notes, tending to equipment, sorting dorsal fin photographs. It was all but impossible with so many people crammed into our tiny trailer. The noise and activity was too distracting, and we were emotionally drained from the constant struggle of trying to accommodate a multitude of diverse personalities and lifestyles in a very small space.

Nortrek, our forty-foot catamaran, alleviated the crowding to some extent. With space to sleep three or four people comfortably, and more if necessary, she became a floating hotel. I always opted to sleep aboard. Rowing out the two hundred yards to where she was moored just offshore of the resort, I felt as if I were going home, putting the noise and crowds and distraction of the resort at arm's length for a while, at least.

I loved the steady, lilting rock of *Nortrek*'s double-hulled body on the water, the sound of wind buzzing through her rigging wires and water rushing past. I loved waking up in the morning to the crowds of welcome swallows, who favored *Nortrek* as a perch. Particularly in the morning, the elegant little black and orange and white birds lined up along the deck safety rails and squeaked and buzzed and trilled to one another in an enthusiastic, charming sort of ramble, all the while eyeing me, squabbling among themselves, and occasionally taking off for a brief, swooping flight before landing again. One pair even tried to build a nest in the starboard hull storage area. Except for the droppings that they left all over the deck, I welcomed their presence. Waking up to their company was far easier for me than facing the crowds of unfamiliar humans on shore in the resort chalets, sipping cappuccino at the tiki hut.

Most of all, I preferred to sleep out on *Nortrek* because by sleeping on the water, to the best of our terrestrial ability, we were truly living among the dolphins. Awakened in the middle of the night by the *pfhoo* of their breathing, I could view the stars and appreciate my place in the universe, floating within the dolphin's realm.

THE MINDS
OF DOLPHINS

After fifteen years, our series of insights into specific aspects of dolphin behavior (many of which have been published in scientific journals) had well documented the minutiae of dolphin daily life: how dolphins catch fish, the composition and fluidity of their groups, the time course of their lives, the social relationships they develop with one another, how and what they communicate. But there had always been a more important, overall question for me about what goes on in dolphins' minds: How smart are they, and how are they smart? And after fifteen years, from the sum of both scientific studies and personal experience, I had begun to get a sense of an answer.

On one occasion, a failing outboard motor provided impetus to ponder some of the similarities and differences between dolphin and human minds. Andrew and I had gone no more than twenty feet from shore, after a difficult start, when the motor quit. While he went back to camp to get some tools and the manual, I struggled to get out the spark plugs for a look.

Within a few minutes three Aussie men were gathered around, helping out. Each had a different idea about the problem, and they were trying to sort out whose plan would take precedence. Tender egos waved like the tentacles of an anemone, outstretched but quick to retract in case of potential injury. "D'ya have fuel in the tank?" asked number one. (Brilliant, Sherlock, yes, I've checked.) "Prob'ly a bit o' grit in the carby," suggests number two. "Sounds like a spark problem to me," says number three. I could tell that at least two of the three knew less than I did about outboard motors, but I didn't dare to tell them so. Their intentions were kind, and men can be sensitive in such situations. One wanted to dump out the fuel tank and clean it with alcohol, the other wanted to take the carburetor apart. Number three was willing to let me continue on the spark plug removal trajectory but insisted on holding the wrench himself.

As this scene unfolded, Nicky approached with a wary look in her eye. She wanted to see if we had any fish for her but didn't want to be bothered. When nobody made a move toward her, she edged closer and parked herself directly behind the motor where she could view the show. She tilted her head sideways and eyed each of us in turn. Then she tilted her head to the other side for a slightly different perspective, taking in the pile of tools and the uncovered motor.

The voices of the three men, edgy and insistent, interrupting one another, faded into the background as I watched Nicky watching us. Apparently sensing my attention, she glanced at me, caught my eye briefly, then moved closer to the boat as if to get a better look, making it clear that she was not interested in me. I so wished that I could know what was going on in her mind.

Equipped with my knowledge of Nicky's day-to-day life, I suppose I am at least in a better position to *guess* at what she *might* think and feel. My guesses may be just me projecting my own ways of thinking onto another being, creating scenarios about Nicky's inner life that are completely wrong, an artifact of my own brain's incapacity to imagine anything all that different from itself. But I hope with some justification that Nicky and I have enough in common that my anthropomorphizing is not too far off the mark. And I can make sure that any ideas about the dolphin mind make evolutionary sense. We know that the dolphin mind has evolved over millions of years, and we know something of the environment in which it evolved. So it should be possible to make some educated guesses.

Nicky watched intently as we fussed with the motor. It was rare to see her so attentive to human affairs other than the feeding of fish. She tossed her head when one of the men approached her, obviously not wanting to be distracted. Why was she so interested in the motor? Could she possibly make any sense of what we were doing?

Though Nicky was obviously curious about our motor repairs, I doubt she has much ability to grasp how outboard motors work. One thing we definitely do not share with dolphins is our propensity to make things. It takes just a quick glance around to realize the importance of human-made objects in our lives. Virtually every human culture has its own, sometimes quite overwhelming, material culture. With our opposable thumbs and fine manual dexterity, we gradually evolved the capacity to make more and more sophisticated things, from stone tools to carved figurines to houses, cars, and microscopes, to satellites and nuclear weapons.

Our propensity for making things permits us to do all sorts of things that would be impossible otherwise. Clothing and houses allow us to populate inhospitable climates; spears, bows and arrows, guns, and dynamite have enabled us to vastly overpower our prey and to engage in vicious warfare with each other. Human material technology has enabled us to dominate our ecology rather than be dominated by it, and some argue that this is the hallmark of human intellect.

But what of dolphin intelligence? Nicky has had ample exposure to outboard motors, but dolphins and their ancestors lack the ability to manipulate objects with any degree of dexterity and simply never evolved the capacity for material technology. There is no precedent, no evolutionary incentive, for mechanically inclined dolphins.

I would bet, however, that Nicky does understand something of the interpersonal dynamics going on between the three men and myself as we stand gathered around the outboard motor. Her mind is a social mind, her intellectual skills lie in the realm of relationships, politics, social interaction. Even though we may be most impressed by human technological innovations, our own intelligence also probably evolved as a social tool.

Nicholas Humphries, in a now famous paper entitled "The Social Function of Intellect," was one of the first to suggest that the need to navigate complex social relationships provided the necessary impetus toward evolving intelligence in humans and, albeit to a lesser extent, in other animals. Robin Dunbar, a primatologist, showed that primate species that live in large social groups and whose individuals have ongoing and diverse relationships have the largest brains. A simple but telling correlation. The more diverse sorts of social relationships one must navigate, the more brainpower is required.

Animals may form social groups initially for various reasons, probably the most significant and ubiquitous of which is protection from predators. Members of a school of fish, a cloud of mosquitoes, or a herd of wildebeest all benefit from forming large groups because they are less likely to be singled out by a predator. Biologists refer to such groups as "selfish herds," because the members are acting purely in their own interests at all times. They don't actually cooperate with one another or try actively to protect one another. Protection is simply a by-product of forming large groups.

It is in those species where group members do actively cooperate with one another that things get more interesting. The benefits of cooperating may be enormous—working together to deter predators or to secure prey that single individuals could not capture on their own—but there are also costs. If you live in a group, you are more likely to have to share your food and more likely to have to compromise when it comes to deciding where you will go from day to day. In fact, group living requires constant compromise and negotiation, because not all individual members of a group have common interests and concerns. The interests of males differ from those of females; those of females with infants differ from those of females without infants; those of high-ranking, dominant individuals differ from those of low-ranking individuals; and so on. Yet in order to reap the all-important benefits of group living, somehow, everyone must get along.

In this situation it is easy to see the advantages of a little political savvy, some diplomatic skills, the ability to outsmart and manipulate others, to envision the consequences of your behavior before you undertake an action, to empathize with others and assess what they may or may not know and how they feel, to deceive others when it is advantageous to do so, and perhaps even to reflect on one's "self" in relation to other group members.

These and many related mental attributes are referred to collectively as "social cognition," and they ring clear and true in their familiarity to us. Humans are consummate social problem solvers. Indeed, we may devote some of our brain space to things technological, but much, maybe even most, of what goes on in our minds, directly or indirectly, in part or in entirety, is social. We compete for status and attention, we tell each other just those things that serve our purposes, excluding the parts that do not, all the while trying to assess what the other party is thinking. We worry constantly about how others perceive us—as generous or stingy, properly moral or immoral, dressed appropriately or not. We sweat over whether someone else does or does not know some "dirty little secret" or did or did not tell a third party. We gossip about each other and spend countless hours engaged in imaginary social encounters.

What strikes me as the most plausible and complete scenario for how human social smarts evolved comes from Richard D. Alexander, who maintains that humans first began to live in social groups that cooperated to defend themselves from predators and to hunt. This sort of cooperation enabled humans to overcome many of the ecological challenges they were faced with, and ultimately, the hostile forces of nature (starvation, predation, disease and inclement weather, for example) became secondary concerns. The primary threat people faced then was other people, either members of their own group or members of other groups.

Humans were forced to cooperate ever more intensely and in ever more complex ways in order to compete against other cooperating groups of people: person against person, family against family, group against group, village against village, community against community, nation against nation. Competitors had at their disposal the very same tools: the social skills to garner support, outsmart, manipulate, deceive, work together, enforce rules of conduct, keep track of favors and insults, play out social scenarios in their heads, and so forth. Matching wits against wits, an escalating, intraspecies "arms race" was established that led to an ever more sophisticated and highly social intelligence. The notion that humans are their own worst enemies may be disheartening, but it nonetheless rings true. Just look at the number of people who die at the hands of their fellow humans in horrific genocides, wars, ethnic conflicts, and random murders.

Could the same process have led to the evolution of dolphin intelligence? Dolphins are both predator and prey—they hunt for the fish that they eat, but they are also hunted ruthlessly by large sharks and killer whales—and the benefits of group living for dolphins probably include cooperative hunting and protection from predators.

I have mentioned some examples of cooperation during hunting from our research in Shark Bay, and there are others, notably from Bernd and Melany Wursig's studies of dusky dolphins off Argentina. The Wursigs watched as the dusky dolphins worked together to herd a large school of anchovy into a tight ball, then took turns passing through the middle of the ball, gobbling up fish and swimming around the perimeter of the ball to keep the fish from escaping. There are also reports of dolphins cooperating to fend off sharks, and certainly with more eyes and ears, the chances of detecting a predator before it can attack are better.

But having watched the intensity and complexity of dolphins' social interactions, I am convinced that this is the driving force behind dolphin intelligence. First of all, because dolphin groups are fluid, changing repeatedly as members join and depart, an individual dolphin comes into contact with a large number of other dolphins in the course of a day. He or she has both a few steady companions and a constantly shifting social milieu. Given the correlation between brain size on the one hand and group size and complexity on the other that Robin Dunbar demonstrated for primates, dolphins *should* have the very large brains they do in fact possess.

It is not just the form of dolphin groups, but what they do within their groups that is most impressive, though difficult to convey. I can recount individual stories and case studies for hundreds and hundreds of pages, and it is the combined effect of many, many different observations, some more complete than others, that has allowed me to trust my gut feelings to shape my thinking about dolphin social intelligence.

There are virtually no systematic studies of dolphin social intelligence, by our research team or any other. We can see the outcomes of behavior, but we know nothing about what went on in the dolphins' mind before, during, or after. We can infer conscious intent or awareness, but almost always the possibility remains that a more conservative, more parsimonious interpretation—an explanation based less on conscious awareness, intent, and forethought and more on simple stimulus-response sequences—*could* be more accurate.

All qualifications aside, there are things we have observed over the years that have certainly shed light on the range of dolphin social and mental life. For example, males who choose to cooperate with one another apparently go through a lengthy process of assessing each other's worth before they even begin to form an alliance. Size and health are likely factors taken into consideration, but do they also have a sense of characteristics like intelligence, loyalty, fairness, reputation, and status in the community? I bet they do and that they can assess such characteristics both through direct personal experience and by monitoring the opinions and interactions of others.

Over the years, we watched one young male, Pointer, who began as a member of an alliance with two other young males, Lucky and Lodent. Whenever the older male alliances in the area encountered Lucky, Pointer, and Lodent, they seemed to single out Pointer as the object of their sex play. They would chase him around, often with erections, splashing and jawing and ganging up on the poor kid. The males Realnotch and Hi, whom we suspected might be the most dominant alliance in Red Cliff Bay, often joined in the fun, harassing Pointer. It was difficult to tell whether Pointer enjoyed these attentions—they were often pretty rambunctious and rough. But somehow he always seemed to end up in the middle of it, when he probably could have taken off or simply avoided the older males if he wanted.

After some years, when Pointer had fully matured, he abandoned his former alliance partners, Lucky and Lodent (who disappeared), and became the newest and by far the youngest member of the Realnotch and Hi alliance (which now also contained Bottomhook, with whom Pointer paired). Pointer's willingness to ingratiate himself with the older males began to seem like a testing ground and his later status an indication that he had proven himself worthy.

Once allied, males are under constant strain to keep their relationships together. There are many sources of conflict that require constant negotiation. Among the most obvious is the problem of who is going to get to mate when the males herd a female. If one male feels slighted, he could "defect" and seek another alliance. If he has few options available, however, he might have to tolerate his partner's transgressions, and his partner might be well aware of that fact and take advantage of it. What if another male seeks to join the alliance or to ally himself with one member of an already allied pair, or if two members of a triplet become more cozy, leaving the third on the outs?

When Snubnose, Bibi, and Sicklefin first solidified their alliance, they seemed to have a roughly equal division of herding opportunities. They took turns being the odd man out. As the years went by, Snubnose was the odd man out more and more often. He seemed to be falling out of favor somehow, while Sicklefin and Bibi grew closer and closer. But Snubnose didn't sit around and sulk about it. He began to work on his relationships with the males Wave and Shave, who had always been tight with Snubnose, Bibi, and Sicklefin. Snubnose often herded with Wave or Shave instead of his alliance partners.

Males have to contend not only with relationships with their alliance partners, but also with other alliances. If alliance A assists another alliance, B, in taking a female from a third alliance, C, is B now obligated to assist A? Dolphins, like people, are probably good score keepers. An important part of being a smart, social, alliance-forming animal is keeping careful track of who helped out whom and when, who owes favors, and just how big that favor should be.

D olphins also may share with us one of the most potent of all abilities for a social mind: the capacity to imagine social scenarios in their minds before actually engaging in them. By doing so, they can consider the actions and reactions of different players given various contingencies and choose among the most favorable outcomes. Social scenario building is such a pervasive part of our own behavior that we hardly even notice it. We think in conversations with other people, imagining their responses based on our elaborate understanding of their personalities and inferences about their feelings.

Dolphins sometimes behave as if they have thought through a scenario in their minds. Take the example of Trips and Bite, when they came into Monkey Mia to assess the attentions Snubnose, Bibi, and Sicklefin were lavishing upon Holeyfin. They went straight out to find Cetus, their third alliance partner, and Realnotch and Hi, their second-order alliance partners. All five males together then came back to Monkey Mia and stole Holeyfin away from Snubnose, Bibi, and Sicklefin. It is hard to imagine that Trips and Bite did *not* have a scenario in mind when they went off to recruit assistance.

A related ability, the ability to empathize—to understand what another individual might be feeling and thinking—is another key facility for a social mind. There is a burgeoning literature in animal behavior on "theory of mind," which examines just what inferences an animal makes about the mental states of other individuals. For example, if a chimpanzee sees another chimpanzee watching someone hide some bananas, does he now infer that the other chimpanzee knows where the bananas are? Does he assume that the other chimpanzee has a mind not unlike his own? Does he make use of what cues are available (gaze, for example) to gather information about the other chimp's knowledge and state of mind? Can he empathize with the mind of another?

To demonstrate unequivocally that an animal has a theory of mind is notoriously difficult. We know we have it ourselves; we are constantly monitoring the behavior of others and inferring what goes on in their heads. And, using language, we can query one another to discover that others do the same. Dolphins behave as if they are empathetic. They are one of the few creatures that will come to the aid of an injured group member, supporting the injured party near the water surface to ensure that he or she does not drown. The same caregiving behavior is even extended to other species, including humans.

One of the most impressive features of human social behavior is that we develop and adhere to moral systems or rules to govern our behavior. Such rules, many of which we take so for granted that we are unaware of them, stem from deeply ingrained cultural and religious indoctrination. ("Do unto others as you would have them do unto you," "Covet not thy neighbor's wife" . . .). Even the simple standards we hold one another to in daily life within our homes and communities are part of our moral system. "Keep your elbows off the table, eat with your mouth closed, don't talk back, always say 'please' and 'thank you,' don't fart out loud, if someone gives you a birthday present, you ought to send a thank-you card, maybe even reciprocate."

My parents taught me all these rules when I was a child, and I never really understood their purpose (in some cases I still don't quite get it), but I knew that failure to abide by such conventions could result in anything from mild ostracism, like getting a dirty look for not saying "thank you," to imprisonment for certain types of stealing or lying. Our elaborate systems of government and law are basically ways to implement and enforce the social rules, the moral systems, we deem important.

The dolphins of Shark Bay have frequently made me wonder if they might have some social rules. What made me consider this was observing some dolphins who were very rare visitors to Red Cliff Bay, apparently coming in from afar. They were strangers, yet they were, as far as we could tell, treated without animosity by the residents. Long-distance wanderings (over hundreds of miles) on the part of dolphins have been well documented in places other than Shark Bay. Why is it that these outsiders are not met with greater hostility? Is it possible that there are some social inhibitions against aggression that make it possible for wanderers to pass unhindered through the territories of complete strangers?

My curiosity about dolphin rules was further piqued by a couple of observations involving food. First, we often noticed that when a dolphin caught a large fish, others often gathered around; but even when the captor threw the fish several yards distant (which they apparently do in an attempt to break off the head), no attempt to steal the prize was made by the attending dolphins. It was as if they considered that the fish belonged to the dolphin who captured it, and trying to steal it was simply not an option. The dolphins' lack of aggressive competition over food stood out in stark contrast with the attendant seagulls, who spent most of their time squabbling, trying to steal food from one another.

Then there were a few incidents that gave me special pause to wonder about dolphin manners. One day Andrew and I were attempting to cajole Crookedfin into rolling upside down alongside our boat. We wanted to get a good photograph of her belly speckles for our records. Our method involved tempting her with a fish, holding it just out of reach so that she had to go through some contortions to reach it, hopefully including rolling upside down. The first few times she approached us, Crookedfin tilted but didn't roll all the way over. We had only one fish, so we withheld it, and a minute or two later, when she circled back toward us, we offered it to her once again. This time she tilted way over, revealing much of her belly, but (as so often happens in these situations) the camera wasn't in order and we missed the shot. Again we withheld the fish.

After a couple of more tries, Crookedfin was clearly getting frustrated with us, tossing her head and circling abruptly. Then Puck approached and I decided to toss the fish to Crookedfin, a reward for her patience, before Puck started to interfere. But Crookedfin moved off, and I had to snatch the fish out of the water before Puck could get it. Puck tossed her head and snapped her jaws irritably. Crookedfin approached again, but when I held the fish out underwater toward her, she veered off without taking it.

Then both Crookedfin and Puck approached at once, and I threw the fish out toward them, just to put an end to the frustration. Puck got there first but, to our surprise, merely pushed it away with her rostrum, refusing the fish she had clearly wanted a moment earlier. Crookedfin came toward the fish, but she too refused to take it now, giving it an indignant shove. Puck again approached the fish and pushed it away. They were like two human friends faced with a single remaining cookie on a plate: "No, you take it"; "No thanks, I don't want it, it's yours"; "No, no, no, I insist, really it's yours." And then finally, "Well, okay." After several minutes Crookedfin took the fish and gobbled it down, whereupon Puck approached her, butting her shoulder playfully against her mother.

A rare observation like that one demands some serious thinking about dolphin minds, but I often found that even as I was watching the dolphins do something I considered commonplace—not terribly informative or especially interesting scientifically—I experienced a sudden shift in perspective that caused me to startle at the parallels between what I was observing in the dolphins and what I know of human nature.

I was getting ready for a purely recreational swim with Puck, who was waiting for me at the shore. Sleek and glistening, she had partially stranded herself on the sandy beach directly in front of me. Though her neck is not very flexible, she strained to lift her head above the plane of her body, watching me as I stumbled into flippers and mask and snorkel. Flapping awkwardly down the beach toward her, I imagined an amused glint in those brown eyes. She pushed off the beach with her pectoral fins as I entered the water and cleared my mask. This was going to be a treat. So often I had gone to the considerable hassle of donning equipment and braving the chilly waters only to be completely ignored by the dolphins. This time Puck was clearly looking forward to my company and I to hers. Nicky and Holeyfin were somewhere nearby as well.

Puck immediately slid alongside me, placing her soft brown eye up to my mask. A tiny bit of white was showing around the corners of her eye, a sign that she was relaxed. I was face-to-face with a wild dolphin. She whistled, and the corners of her crescent-shaped blowhole puckered and wriggled a little with each whistle. It is not easy to make sounds underwater, but I did my best to imitate her with as dolphinlike a greeting as I could achieve.

I stroked Puck's side and maneuvered into deeper water where we would both be able to move more freely. Nicky joined us. Nicky and Puck, both adolescents at the time, reminded me of some human girls I have known. Their relationship was close but fraught with jealousies and little grudges over one thing or another. This played out as the two begin to vie for my attention. I held out a piece of seagrass in one hand, and both rushed to grab it. Nicky, in typical style, bumped Puck out of the way and took the grass. Puck shook her open jaw at Nicky irritably and lunged to grab it back from Nicky, but she failed and instead approached me again.

Nicky, seeing that the weed was no longer the focus of Puck's desire, dropped it disinterestedly. Puck, outgoing and affectionate, laid herself across my lap. She seemed huge, her body hard, like taut rubber, but warm and, at the very surface, softly yielding to the touch. I stroked her, and she tilted her pure white belly toward me. Under her "armpits" and her chin she had several tiny creases, lined in delicate pink. I tickled her there and she wriggled, but she tolerated it and did not move away.

Nicky then tried to push in between Puck and me, but Puck prevented it. Instead Nicky circled around to the other side and offered me a piece of seagrass. It was not that she wanted so badly to play with me, I knew, but that she was trying to distract my attention from Puck. Puck tossed her head at Nicky and placed herself between us. Stroking down the length of Puck's belly, my hand approached her genital slit. She thrust forward as though to increase the level of contact, making a point, I think, to Nicky rather than me. I obliged, but Nicky lunged abruptly at Puck with jaw open, scraping her teeth against Puck's forehead. The two of them rolled and tumbled together amid a cacophony of squeals and grunts. I tried to get out of the way. No point risking a knock from those powerful tail flukes or having three hundred pounds of solid muscle bump into me.

When they settled down a bit, I could see that Puck now had a bit of seagrass in her beak (the same piece?). Nicky rushed up to grab it from her, but Puck tilted away and evaded her, so Nicky approached me instead and was uncharacteristically tolerant of my stroking and touching, lingering across my lap just as Puck had been. But I could see that she was largely oblivious of me. Instead her eyes were following Puck even as my hand moved gently down her side.

At bottom, their bickering had little to do with me. They were just like a couple of human children (or even adults) bickering over a favored toy. Nicky and Puck were using me and the seagrass as tools with which to sort out their own business. I gladly complied in exchange for the privilege of watching. As the bickering, interference, and rivalry went on and on, I was struck by the powerful realization that these dolphins spend most of their time and mental energy sorting out their relationships.

Obviously there are many aspects of dolphin existence that are very different from our own. Besides the huge differences in our physical bodies and sensory abilities, dolphins live in the ocean while we live on land. They rely completely on hunting for their food while humans have developed agriculture and husbandry. Dolphins are subject to serious predation by sharks while humans now have no real predators. While we form monogamous pairs and human fathers contribute to taking care of their offspring, dolphins mate promiscuously and probably don't know who fathered whom. Dolphins and humans have evolved for millions of years in vastly different environments and from very different ancestries. A list of differences could go on and on. So in taking account of all my ideas and impressions about dolphin intelligence, I am amazed at how much we actually seem to have in common. Like us, dolphins spend much of their brainpower keeping track of who does what with whom, engaging in rivalries and social politics, figuring out what others might be thinking, competing and cooperating in complicated, multileveled alliances. Like us, their minds are on each other.

So what is it that we see when we look into the eye of a dolphin and are struck immediately and powerfully with a sense that theirs is an intelligence of extraordinary measure? What we see is an eye that is not obsessively hopeful and approving like that of a dog, or absently self-absorbed like that of a cat, or fearfully distracted like that of a bird. Rather, it is an eye that seems somehow familiar even as it defies description—one that appraises, reserves judgment, watches intently, weighs and considers, infers and understands. It is an eye that can strategize, empathize, and, perhaps above all, recognize when its gaze is met by another, similarly complicated and sentient being.

DOLPHIN

CONSERVATION

One April afternoon during our study of mother-infant communication, Janet and I did a follow of Peglet, Square's six-month-old baby. The dolphins spent most of the afternoon resting, with Peglet tucked up close against Square's side in baby position. They stayed mostly in shallow water over the flats north of Monkey Mia, slow traveling with occasional bouts of snagging. Square poked around in the weeds on the bottom, flushing out some small fish which she pursued lackadaisically, but she seemed more interested in resting than hunting.

After several restful hours, Yogi and Smoky joined. Smoky's arrival represented an opportunity to play, and Peglet was instantly transformed from a listless bump against Square's side into a highly animated bundle of cuteness. The two babies splashed around and chased each other, tumbling and chasing, jawing and butting each other playfully while their mothers maintained a holding pattern nearby, floating side by side with eyes half-shut, waiting for the kids to catch up. Peglet and Smoky took an interest in circling our boat. Tilting sideways to look up at us, Peglet held her pectoral fin pressed against Smoky's side as if for reassurance.

Suddenly I heard an intense, screamlike whistle through the hydrophone. Then Peglet broke through the water surface, leaping straight up toward the sky, pausing momentarily before she came crashing back into the water. This was not a traveling leap, which would be low and directional, intended to cover distance. I heard more intense whistling, loud and quavering, as though produced under such great pressure that air was released along with the whistle. Again Peglet leaped, now with Square at her side. Mother and baby took off, leaping several more times in tandem. Confused by this sudden drama, we followed them in hopes of figuring out what it was about. After a few minutes we caught a glimpse of something trailing behind Peglet. A ball of tangled fishing line trailed off her dorsal fin. More line was wrapped around her dorsal fin, probably lodged somewhere on her body by a hook. I felt my heart drop into my gut.

Square and Peglet were swimming at top speed, side by side. Each time they took a breath, they were either porpoising or doing traveling leaps together and breathing hard from the exertion. Janet and I had to yell back and forth over the thrumming of the motor as we tried to keep up with them. They swam erratically, changing direction frequently and for no obvious reason as if they thought the ball of fishing line were pursuing them and they were trying to evade it. Then two other dolphins joined Square and Peglet, leaping alongside, keeping pace with them at a few yards' distance. It was Surprise and Puck. The entire group turned east and continued racing along.

The tangled ball of fishing line dragged along behind Peglet, probably gouging into her flesh and causing the hook to lodge deeper. After keeping pace with Square and Peglet for about ten minutes, Surprise and Puck slowed and broke away. Perhaps seeing Square's and Peglet's fear had frightened them, and, not knowing the cause of the fear, they had taken the safest course: flee in the same direction.

Square and Peglet continued racing around the bay. Three other dolphins appeared, and they also paced alongside Square and Peglet at top speed, then broke away after about ten minutes, leaving Square and Peglet racing out to the north on their own.

Janet and I struggled to figure out a way to convince Peglet to approach us so that we could untangle the line from her. If only we could convey to her that we could help, that we could do things with our hands that they are incapable of doing with their flippers. Outrageous though that may seem, I had reason to suspect that dolphins might be capable of that sort of thinking. Wilf Mason had described an incident where a strange adult dolphin who was clearly in trouble came into the shallows at Monkey Mia and approached him. She had a large fishing hook lodged in her mouth. This dolphin, unaccustomed to human contact, had permitted Wilf to remove the hook with a pair of pliers. This is all the more remarkable because it must have hurt like hell to have the hook dislodged, yet she somehow understood that Wilf was helping her, that the pain was ultimately necessary, and that she would be better off in the long run if she tolerated it. I'd heard other similar and equally remarkable stories of dolphins seeking help from humans.

But still, Square or Peglet were unlikely to voluntarily seek our help. Could we capture them using a net, with the help of the Monkey Mia rangers? This was also difficult to envision and potentially highly stressful for Square and Peglet. We followed along helplessly as they continued to flee in terror from the danger dragging along behind. They had been traveling at top speed for well over half an hour. Then we crossed over the flats north of Monkey Mia and hit the edge of the deep-water channel, where the water grew suddenly dark and waves smacked against the sides of the boat and tossed us off course. It would be dangerous to keep heading offshore at this hour in these conditions. There was nothing we could do anyway. Janet had continued to collect follow data throughout the whole event, but now we were losing track of the dolphins too often, and it was time to call it quits.

We turned to shore and headed grimly back to Monkey Mia, wondering if square and Peglet would continue to race around as long as the fishing line stayed lodged on Peglet's fin, perhaps becoming dangerously exhausted. Would the line dig into her flesh and cause some horrible injury or get tangled onto other parts of her body, Square's body, or something else, perhaps drowning Peglet? The prospect of her death as a result of someone carelessly discarding a tangle of fishing line was infuriating.

Back at camp, worry cast a pall over our evening chores and dinnertime. We vowed to get up early in the morning and try to find Peglet as soon as we could, but at dawn the wind was raging out of the southeast and it would be too rough to search for them. We waited all morning and into the afternoon. Still the wind blasted. Finally, in frustration, we decided to hop on the boat anyway. At least we could search the waters right near Monkey Mia. Square and Peglet didn't usually frequent these waters, so it felt a bit like searching for lost keys under a street lamp because that was where the light was; but we had to do something. As if they knew we would be looking for them, Square and Peglet were the first dolphins we encountered, not half a mile from shore. The fishing line was gone. Peglet's fin had a fresh slice where it had cut into her, but she was okay. We clapped and yipped and whistled our delight to Square and Peglet, then headed back to Monkey Mia, tremendously relieved, at least for the moment.

Peglet was not the first dolphin to get tangled in fishing line at Monkey Mia. No doubt there was a lot of discarded line floating around, as people often fished off the jetty. Babies seemed especially vulnerable; perhaps their curiosity led them to explore more, or perhaps they had not yet learned about the dangers. But the Shark Bay dolphins are relatively lucky. Their problems with monofilament line and fishing nets and hooks are negligible compared with those faced by dolphins elsewhere in the world.

The problems are so serious that I have a difficult time fathoming the statistics reporting how many dolphins die each year or each day or each minute as a result of human fishing practices. The highly publicized example is the fishing industry. In order to catch tuna, fishermen encircle herds of dolphins with whom the tuna associate. Until just a few years ago, when dolphin mortality declined owing to new restrictions and technologies, according to IATTC (InterAmerican Tropical Tuna Commission) estimates, over one lakh dolphins were killed each year in the process. Yet even now, an estimated three thousand to five thousand dolphins are killed each year by the tuna fishery. Thousands of dolphins, each one like Nicky, Puck, Holeyfin, Surprise, Square, Squarelet, Peglet, Snubnose, Bibi, Sicklefin—all those bright dolphin personalities extinguished in the name of providing tuna-fish sandwiches.

The tuna fishery is just one offender. Worldwide, it is estimated that each year fishing boats pull up twenty-seven million metric tons of "nontarget" marine life that is thrown overboard, dead or dying. This "bycatch," which includes dolphins and other marine mammals, represents about one-quarter of what is caught. In reality, anywhere fishing nets made of tough, durable monofilament plastics are used to capture fish, dolphins are in trouble. That means virtually everywhere in the oceans and rivers inhabited by dolphins.

Entanglement is not the only human-caused danger threatening dolphins. In some places dolphins are hunted purposefully for food or, as was the case until recently off Chile, for use as crab bait. Some dolphin populations have declined because the fish species they depend on for food have been overfished by humans. And river-dwelling dolphin species are among the most highly endangered, largely because so many major riverways the world over feature hydroelectric power dams all along their length, making them impassable to dolphins. The Yangtze River in China is a case in point. The Baiji dolphin used to call the Yangtze home, but dams, vast quantities of boat traffic, dolphin-unfriendly fishing practices, and pollution have virtually wiped out this species. Current estimates indicate that there may be fewer than one hundred Baiji remaining, and their future looks grim indeed.

Pollution, in its many insidious forms, is probably the greatest threat to dolphins worldwide. Because dolphins are predators at the top of the food chain, they are in the unfortunate position of bioaccumulating toxins. Bioaccumulation happens when a tiny organism, say, a small fish, accumulates toxins (like DDT or PCBs) into its tissues through its diet. Some larger fish then comes along and eats many of those small, somewhat toxic fish. The larger fish now has accumulated in its tissues all of the toxins from all of those tiny fish it consumed. Now an even larger fish comes along and eats the big fish, and many others of its kind, accumulating the accumulated toxins. By the time we get to a predator like a dolphin, which feeds on fish that ate lots of other fish that ate lots of other fish, the amount of accumulated poison can be truly alarming.

Researchers have found extremely high levels of organochlorine contamination in the tissues of marine mammals, including dolphins. We know that these poisons interfere with reproductive endocrinology, immune system functioning, and even cognitive development. Even worse, these contaminants are passed on from generation to generation through nursing. The toxins are fat soluble, so when a mother metabolizes body fat to produce milk for her offspring, the toxins end up in her milk. Babies therefore begin life with a huge load of toxins before they even begin accumulating through their adult diet. The result is an exponential increase in toxin load from one generation to the next.

Climate change, caused by a combination of ozone depletion, burning of hydrocarbons, and deforestation, is another serious threat to dolphins and other marine mammals. It is difficult to predict exactly how dolphins will be impacted, but entire ecosystems, including dolphins, will undoubtedly be affected by rising temperatures, decreasing salinity, rising water levels, and other forecasted changes.

Shark Bay is one of the most remote places on the face of earth. The beaches are still relatively litter free, the air and water relatively clean. Yet even there pollution has taken its toll. In February 1989, during an interim visit to the United States, I got a phone call from Richard, from Monkey Mia: "I have some bad news." I could tell by the tone of his voice that a dolphin had died.

"What?" I asked.

"Snubnose, Bibi, and Sicklefin are gone."

"What do you mean, 'gone'?"

"They haven't been seen here at Monkey Mia or offshore in a week. Chances are that they are dead; otherwise they would be coming in."

It was not one dolphin, but all three of the males. I hung up the phone feeling stunned. After all that we had learned from those three males, I couldn't even imagine Monkey Mia without them.

Then an even greater shock came a day or so later when Richard called again, his voice shaky and solemn. "Rachel, Holly is dead, too . . . and Nipper [Nicky's firstborn]." When it had been just the adult males, we could still harbor some hope that they had just gone off somewhere else for a while and would return. But not Holly and the baby. They were too young to be taking off on their own. Our dolphins were dying.

The death of Holly was by far the most difficult for me to accept. She and I had enjoyed so many extraordinary playtimes. Her little spirit seemed so bright and playful and alive that I could not accept it had been extinguished. She was so young and full of the potential of a long and interesting dolphin life. We would never get to see her mature, have babies of her own, be a mother, grow old. I would never again play sea-grass games with her or swim by her side.

The deaths were caused by pollution. The tides had been extreme, as they tend to be at that time of year. The previous year a new toilet block had been put in place to provide facilities for the growing numbers of tourists. The septic system was installed only a few feet from the high-tide line. With unusually heavy use for that time of year, the tanks had been literally flushing out into the bay. After the fact, water tests revealed very high levels of bacterial contamination in the nearshore waters of Monkey Mia. Though some argued that there was no connection, it seemed unlikely to be just coincidence. What a sad irony that the dolphins were literally poisoned by the excrement of their admirers.

The deaths happened around the time Monkey Mia was in transition from Wilf and Hazel's camp to the modern resort. We had fought hard with Wilf and Hazel to get restrictions placed on the feeding of the dolphins, and then we had gone on to fight for other controls that we believed were important to the continued well-being of the dolphins. In spite of our initial fears about the development at Monkey Mia, we found the new management to be quite receptive to our input on the dolphin's welfare. They seemed to recognize that the dolphins were the draw and that keeping them alive, healthy and continuing to visit Monkey Mia, was very much in their interests.

We made recommendations about all sorts of things: what and how to permit tourists to feed and interact with the dolphins, restrictions on the speed of travel for boats in nearshore waters, limits on the use of pesticides and chemical fertilizers that could run off into the water, ongoing monitoring of the water quality, and limitations on concessions for the rental of joyriding devices like jet skis and even sailboards (under a strong wind, a sailboard can get going up to speeds of sixty miles per hour, and they are relatively quiet; a resting or inexperienced dolphin could easily get hit, as has happened in other places where dolphins and sailboarders mix).

Over the years we were able to collect enough data on infant survivorship to demonstrate that the Monkey Mia dolphins were experiencing even higher levels of infant mortality than were the dolphins who did not visit Monkey Mia. We worried that the babies were not learning how to behave like proper dolphins because they were spending too much time with their mothers getting free handouts at Monkey Mia instead of foraging out in the wilds of the bay.

Long gone are the free and easy days of our early Monkey Mia experience. Now the resort and the dolphins are a "managed resource": the only way to cope with the vast numbers of visitors who want to see the dolphins, feed them, touch them, and get a picture taken. A staff of full-time employees works to control the crowds and serve as interpreters, answering the questions of visitors.

Though these changes have been saddening to me in some ways, I have also felt heartened by the fact that so many people take an interest in the dolphins, an interest that is great enough to motivate people to travel all the way to Shark Bay. Their appeal is ultimately the dolphins' greatest hope for preservation.

Having visited these dolphins on and off for so many years, I suppose I have come to take them for granted in some regards. Occasionally I step back and realize what a tremendous thrill it is for most visitors to wade into the water and touch a wild dolphin. I recall the occasion of my own first contact with Holeyfin many years ago, the surprising warmth of her skin against my hand. And the expression on the old blind woman's face as her husband led her into the water to touch her hand against Puck's side: the fulfillment of a lifelong dream. What better way for people to learn about these creatures and to have their interest and concern sparked?

I believe that the experience is much more powerful because these dolphins are wild. If they were in a tank, there would always be the lingering sentiment that they were simply trained to behave this way or had no other options. The Monkey Mia dolphins force people to meet them halfway, to wade into the water, into their element. Once there, people are met with friendly dolphin faces, but there remains a certain aloofness that bespeaks the fact that these dolphins are wild, have much more going on in their interesting lives than simply entertaining people, and are, for the most part, not dependent upon us.

As I am writing this, we are celebrating the "Year of the Oceans." It began with a petition, signed by hundreds of scientists, policy makers, and concerned citizens, imploring the world's leaders to step up efforts to protect the world's oceans. All are in solemn and ominous agreement that the oceans, heart and lungs of our planet, are dying. Dolphins are just one of the creatures we stand to lose.

I guess we tend to care for individuals whose personalities we can know and with whom we can identify. Therein lies my hope that the Monkey Mia dolphins, who have indeed shared their lives and characters with us, might touch us in a way that will have far-reaching impact on the protection of their kin. If we aren't willing to do what is necessary to save the oceans for the sake of a coral reef, a species of dam-selfish, a manatee, or a banjo ray, or even the future health of our own species, then perhaps Nicky, Puck, Surprise, and their offspring, who now carry on the Monkey Mia tradition of allowing people to touch them, can inspire us.

EPILOGUE

One could easily spend an entire life studying the dolphins and yet continue to make new discoveries. Since dolphin lifetimes are roughly of the same duration as ours, a human lifetime of watching would track only a single generation of dolphins. But as the years passed, I gradually came to realize the toll that camp living was taking on me and decided it was time for a break.

In some ways I felt that understanding the lives of the dolphins had come at some expense of my own. I had watched Nicky and Puck and others grow up, play and hunt, develop their relationships with other dolphins, have offspring of their own, experience losses, and continue with the course of their rich lives. Meanwhile I was being driven to distraction by the crowded living and working conditions and by the feeling of being continually uprooted as I commuted back and forth around the world. I had seen two important relationships ruined by my long absences from the United States. I had worked hard to cultivate friendships in both places but had trouble fitting in quickly and easily to the ever-changing social landscapes I was faced with.

Perhaps it was my advancing maturity, but I began to long for my own space; to unpack my possessions from boxes and put my backpack away in some closet, to have control over with whom I drank my first cup of coffee in the morning, to be able to put something down and know it would be there when next I needed it, to relax quietly with a glass of wine and a book in the evening.

We had come a long way toward making sense of the lives of the Shark Bay dolphins. The more we had learned, the more discoveries we realized still lay ahead. But personally I felt it was time for my own life to take precedence. It was time to settle and put down some roots, cultivate relationships that could endure, finish up the many reports that lay half-completed on my desktop. It was time to slow down on my visits back and forth to Australia and take time to reflect upon my experiences and all that we had learned from the dolphins.

Now it has been a few years since I last spent time with the dolphins of Monkey Mia. My own life has changed dramatically; I married, moved to rural Vermont, put my career on hold after the birth of one child and imminent arrival of a second. Things have changed at Monkey Mia as well. Holeyfin died in the summer of 1997. She was one fine old lady dolphin, but probably feeling her age. The veterinarian who performed an autopsy discovered a stingray spine lodged near her heart, which had caused infection and bleeding. That may have been what finally did her in, but I suspect she was ready. Her long and rich life is probably one of the best-chronicled of all dolphins on the planet.

Nicky, Puck, and Surprise are the current generation of dolphin matriarchs. Puck's daughter Piccolo, Surprise's daughter Shock, and Nicky's daughter Holeykin will probably carry on the Monkey Mia tradition into yet another generation. Meanwhile all three mothers have given birth once again. Previous experience has taught us not to place too much hope in any youngster, at least until he or she has survived the first couple of years; but hopefully they will make it.

With a new family of my own to occupy my time, I have allowed the Monkey Mia days to become a thing of the past, at least for now. A new generation of excellent and motivated students has become involved in the research, and it is nice to think of them experiencing some of the same thrill that I felt in the early days of my tenure there.

Still, when I shut my eyes, I hear the ring of the wedgebill's song, and out of the darkness emerges a dolphin, looking up at me through the water surface, that transparent but profound barrier between us. From such different worlds have we come that we could as easily be aliens from another planet. Yet there is so much that we share. For one thing, we are both curious enough to overcome any fear that might have prevented us from even meeting.

What are you? Who are you? For both of us, the answers to those questions lie in long histories—both evolutionary and personal. Our lives have twisted and turned from birth, through many years of childhood and on into long adulthoods, encompassing great transformations, pleasures and frustrations, good fortune and loss. Though our species bear no resemblance, we have both learned to navigate the tricky waters of our relationships with others, loving, loathing, giving, and taking, trying always to understand what makes our friends and associates tick. We both carry these long and intricate histories into this moment with a common intent. Can we be friends? Hey, you, on the other side there, our hearts and minds are not so different after all!

In my imagination, the image of this dolphin fades, the wedgebill's chime as well. Our meeting has flickered into the past, a fleeting glimpse through a small window. But these sounds and images tug at me, pulling me back to the bay. Soon I may pack up my bags, my husband, and our children and lead us all back Down Under to rekindle those friendships and share with my family the thrill of touching a wild dolphin.

ACKNOWLEDGMENTS

Thanks are due first of all to the dolphins of Monkey Mia, past and present, for allowing us to watch, for putting up with our noisy little boats so patiently, and for showing us so much of their world. My parents, Robert and Rosemary Smolker, instilled in me an appreciation and interest in nature for which I am eternally grateful and without which I would have been lost. Elizabeth Gawain "turned us on" to Monkey Mia. Bernd Wursig was instrumental in getting us on our way and has been a guiding light ever since.

The rangers at Monkey Mia, present and past, deserve thanks for all their assistance over the years and for safeguarding the nearshore realm of the dolphins. Thanks are also due to the Monkey Mia Resort. The Western Australia Museum, especially Darryl Kitchener, provided valuable assistance as did the University of Western Australia, especially Richard Holst. Long Marine Lab at the University of California at Santa Cruz and the Museum of Zoology at the University of Michigan also provided invaluable logistic support stateside.

Major financial assistance over the years came from the National Geographic Society, the Seebie Trust, Gordon and Ann Getty, the National Science Foundation, and the University of Michigan. Thanks are due to Ruth and Ken Musgrave, Irv DeVore, and Richard Wrangham for helping to keep the project afloat.

Several people have served as mentors during my undergraduate and graduate school careers, greatly influencing my thinking and direction. These include Ken Norris, Bernd Wursig, Bob Trivers, Barb Smuts, Irv DeVore, and Richard Alexander.

My co-workers at Monkey Mia, Andrew Richards, Richard Connor, Janet Mann, Amy Samuels, Per Berggren, Mike Heithaus, John Pepper, and Bill Sherwin, all deserve thanks. Many of the experiences relayed in these pages were shared with one or more of the above. Andrew Richards and John Pepper in particular have been colleagues, advisers, and much beloved friends throughout.

My husband, Bernd Heinrich, is an avid biologist and writer, and he encouraged me to write about my experiences, often reminding me that scientific research is largely useless if nobody but a few elite academics knows the results. He has served as a supportive and insightful sounding board during the writing of this book. We met just after my last trip to Monkey Mia, so, oddly, he has never seen the animals and the place that occupied so much of my attention before his time.

Nina Ryan, Cynthia Cannell, and my editors at Doubleday, Sean McDonald and Nan Talese, did much to improve this book manuscript at all stages of development and to shepherd it through the process of publication.

Thanks to you all!